Understanding the Brain

Fetal Alcohol Spectrum Disorders

David Balog

www.HealingtheBrainBook.com

Understanding the Brain

Fetal Alcohol Spectrum Disorders

David Balog
2367 Curry Road
Schenectady NY 12303

518 952-1257
dbalog99@gmail.com

Copyright: 2017 David Balog

About the Author:
David Balog, currently a freelance science/medical writer, served as an editor at the Dana Foundation from 1995-2006. He created, wrote, and edited *The Dana Sourcebook of Brain Science* through four editions. More than 50,000 copies were distributed to elementary schools, middle schools, colleges, and to professionals and the general public. He worked with leading brain scientists and doctors, including Nobel laureates, throughout 12 years at Dana. David Balog is a graduate of Hamilton College, Clinton, New York.

Table of Contents

Dedication..........5

Introduction. Lives Detoured: Fetal Alcohol Spectrum Disorders (FASDs).....6

1. Nurturing Strength and Confidence...........8
2. Exploring the Brain in Words and Pictures...........14
3. Use It or Lose It: Maintaining Brain Health............31
4. The Power of Emotions.........34
5. Wounds that Time Alone Won't Heal: The Biology of Stress..........37
6. Behind the Scenes in the Adolescent Brain.........53
7. A Darker Shade of Blue: Teen Depression..........56
8. Teen Suicide: Death in Life's Springtime.........61
9. TBI and Domestic Violence: Healing a Broken Brain........65
10. Fighting a National Sleep Crisis...........89
11. Minority Stress and LGBT/Q Health......................96
12. "The Best Little Boy in the World".........99
13. Substance Abuse: Saying No Is Very Hard to Do..........110
14. Fetal Alcohol Spectrum Disorders (FASDs)..........132
15. Great Brain Books........142

Appendix I: A Glossary of Key Brain Science Terms......154

Appendix II: Maps of the Brain......173

Appendix III: Learn More About It: Resources on the Brain......178

Every aspect of our lives depends on the normal functioning of our brains. Our education depends on it; the education of our children depends on it; our relationships to our fellow humans depend on it; our hopes and aspirations are all represented in our brain. And all of these human qualities are at risk if something goes wrong with one's brain.

—W. Maxwell Cowan, M.D., Ph.D., Neuroscientist, Educator

Remembering Barbara Rich: Educator, humanist, friend

At a holiday luncheon in New York City in the late 1990s, Dana Foundation chairman David Mahoney was quite proud of his employees. We were making a big difference in the "Decade of the Brain" congressional project, he said.

At Mr. Mahoney's behest, we posed for a photograph in the beautiful garden of the New York Academy of Sciences. I was in the photo, which he graciously signed personally and distributed to all those in the picture. "Remember," he wrote on the photo, "that you were there at the beginning."

Also among those gathered was Barbara Rich, in many ways the emotional and intellectual backbone of this intrepid group of which Mr. Mahoney was so proud.

Barbara was a renaissance woman. Kind, acutely smart, indefatigable, she eventually had a say in all parts of the Foundation, including the financial portfolio. We worked together on mutual projects, supporting each other. I learned a great deal more from her than the other way around. She was always striving to be a better person and gently encouraging me to do the same. She glided when walking with the grace of a dancer (for which she had been trained).

Anyone learning about the brain and concerned about a loved one--or themselves--suffering from depression, substance abuse, suicidal ideation or the myriad of other illnesses that can cripple individuals and minimize human potential, needs to thank Barbara Rich.

With great sadness I learned of her recent passing.

Thank you for your guidance and insights, Barbara. It was a great honor to know you and work alongside you. This book would not be possible without you; therefore I dedicate it to your memory.

--to BR, from DB

Introduction

Lives Detoured

Fetal Alcohol Spectrum Disorders (FASDs)

The brain is a lifelong work in progress.

Development takes place most rapidly before birth, maintains a furious pace in infancy and continues briskly through childhood and adolescence, but never ceases altogether. In the third week of gestation, genes switch on to turn some of the embryo's stem cells — "blank slate" cells with the potential to become any kind of tissue — into neurons and glial support cells. These newly formed cells multiply, migrate and connect with one another, guided by chemical signals into the webwork of brain anatomy. By week seven, primitive forms of the cortex, cerebellum and brainstem are apparent.

> *Fetal Alcohol Spectrum Disorders (FASDs) can affect each person in different ways, and can range from mild to severe.*

Given the complexity of the task, it's astonishing how often everything goes right. The great majority of babies are born with a brain primed for sensing, learning, and developing. Brain growth doesn't end at birth, of course; the prefrontal cortex, for example, will not fully mature until as late as age twenty-two. Yet the overall structure of the brain and nervous system, including virtually all of the cells and a rough outline of the synapses connecting them, is in place at birth.

That is, if everything goes right.

Not everything does in every pregnancy. There is a long list of things that can go wrong that may have subtle, moderate, or devastating effects on the fetal brain. Among them is exposure to alcohol in the mother's womb.

According to the CDC (Centers for Disease Control), Fetal Alcohol Spectrum Disorders (FASDs) refer to the whole range of effects that can happen to a person whose mother drank alcohol during pregnancy. These conditions can affect each person in different ways, and can range from mild to severe.

Healing the Brain: Fetal Alcohol Spectrum Disorders gives readers a view of the remarkable human brain, its capabilities, and its vulnerabilities. A brain compromised by FASD is tragic, preventable, and increasingly yielding to treatments and therapies. Detailed coverage, "Fetal Alcohol Spectrum Disorders (FASDs)," appears in Chapter 14.

Chapter One

Nurturing Strength and Confidence

Trainings in child welfare typically employ the concepts of Erik Erikson and Abraham Maslow. Erikson generated interest and research on human development throughout the lifespan. Maslow, author of the acclaimed book Toward a Psychology of Being, *believed one must satisfy basic needs before progressing to meet higher level growth needs.*

Erikson's Stages of Development

Infancy *(birth to 18 months)* Trust vs. Mistrust
Feeding. Children develop a sense of trust when caregivers provide reliabilty, care, and affection. A lack of this will lead to mistrust.

Early Childhood. *(2 to 3 years)* Autonomy vs. Shame and Doubt
Toilet Training. Children need to develop a sense of personal control over physical skills and a sense of independence. Success leads to feelings of autonomy, failure results in feelings of shame and doubt.

Preschool *(3 to 5 years)* Initiative vs. Guilt
Exploration. Children need to begin asserting control and power over the environment. Success in this stage leads to a sense of purpose. Children who try to exert too much power experience disapproval, resulting in a sense of guilt.

Wikimedia Commons

Erikson and Maslow generated interest in human development through the lifespan.

School Age (6 to 11 years) Industry vs. Inferiority
School. Children need to cope with new social and academic demands. Success leads to a sense of competence, while failure results in feelings of inferiority.

Adolescence (12 to 18 years) Identity vs. Role Confusion
Social Relationships. Teens need to develop a sense of self and personal identity. Success leads to an ability to stay true to yourself, while failure leads to role confusion and a weak sense of self.

Young Adulthood (19 to 40 years) Intimacy vs. Isolation
Relationships. Young adults need to form intimate, loving relationships with other people. Success leads to strong relationships, while failure results in loneliness and isolation.

Middle Adulthood (40 to 65 years) Generativity vs. Stagnation
Work and Parenthood. Adults need to create or nurture things that will outlast them, often by having children or creating a positive change that benefits other people. Success leads to feelings of usefulness and accomplishment, while failure results in shallow involvement in the world.

Maturity (65 to death) Ego Integrity vs. Despair Reflection on Life. Older adults need to look back on life and feel a sense of fulfillment. Success at this stage leads to feelings of wisdom, while failure results in regret, bitterness, and despair.

Maslow: The 12 Characteristics of a Self-Actualized Person

Abraham Maslow describes the good life as one directed towards self-actualization, the pinnacle need. Self-actualization occurs when you maximize your potential, doing the best that you are capable of doing. Maslow studied individuals whom he believed to be self-actualized, including Abraham Lincoln, Thomas Jefferson, and Albert Einstein, to derive the common characteristics of the self-actualized person. Here are a selection of the most important characteristics, from his book *Motivation and Personality*:

1) Self-actualized people embrace the unknown and the ambiguous.
They are not threatened or afraid of it; instead, they accept it, are comfortable with it and are often attracted by it. They do not cling to the familiar. Maslow quotes Einstein: "The most beautiful thing we can experience is the mysterious."

2) They accept themselves, together with all their flaws.
She perceives herself as she is, and not as she would prefer herself to be. With a high level of self-acceptance, she lacks defensiveness, pose or artificiality. Eventually, shortcomings come to be seen not as shortcomings at all, but simply as neutral personal characteristics. "They can accept their own human nature in the stoic style, with all its shortcomings, with all its discrepancies from the ideal image without feeling real concern [...] One does not complain about water because it is wet, or about rocks because they are hard [...] simply noting and observing what is the case, without either arguing the matter or demanding that it be otherwise."

Nonetheless, while self-actualized people are accepting of shortcomings that are immutable, they do feel ashamed or regretful about changeable deficits and bad habits.

3) They prioritize and enjoy the journey, not just the destination.
"[They] often [regard] as ends in themselves many experiences and activities that are, for other people, only means. Our subjects are somewhat more likely to appreciate for its own sake, and in an absolute way, the doing itself; they can often enjoy for its own sake the getting to some place as well as the arriving. It is occasionally possible for them to make out of the most trivial and routine activity an intrinsically enjoyable game or dance or play."

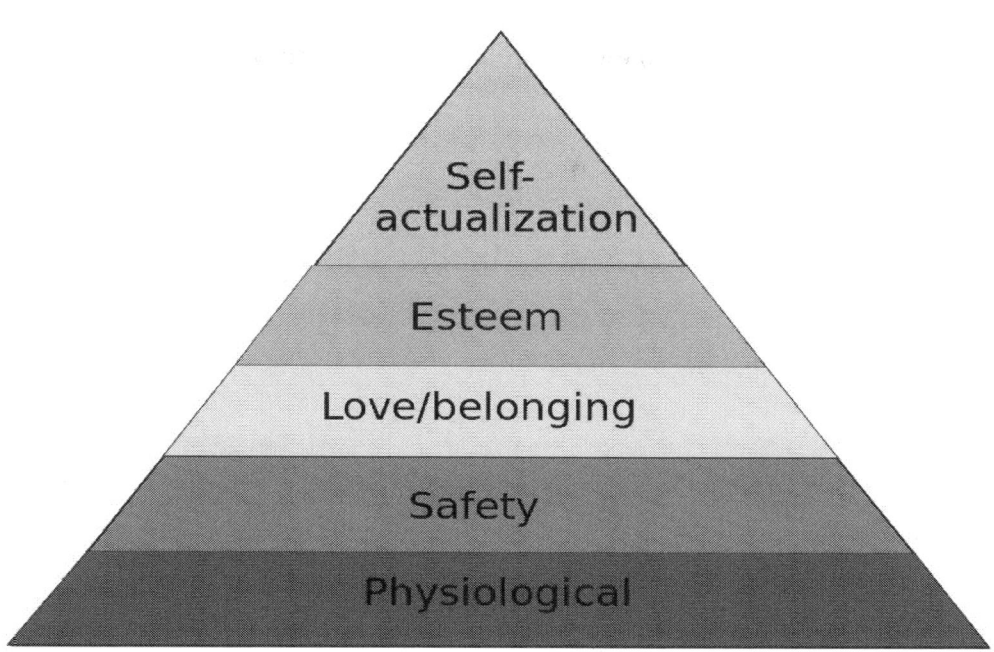

Wikimedia Commons
Maslow's Hierarchy of Needs holds that self-actualized people are motivated by growth and development.

4) While they are inherently unconventional, they do not seek to shock or disturb.
Unlike the average rebel, the self-actualized person recognizes:
"… the world of people in which he lives could not understand or accept [his unconventionality], and since he has no wish to hurt them or to fight with them over every triviality, he will go through the ceremonies and rituals of convention with a good-humored shrug and with the best possible grace [… Self-actualized people would] usually behave in a conventional fashion simply because no great issues are involved or because they know people will be hurt or embarrassed by any other kind of behavior."

5) They are motivated by growth, not by the satisfaction of needs.
While most people are still struggling in the lower rungs of the 'Hierarchy of Needs,' the self-actualized person is focused on personal growth. "Our subjects no longer strive in the ordinary sense, but rather develop. They attempt to grow to perfection and to develop more and more fully in their own style. The motivation of ordinary men is a striving for the basic need gratifications that they lack."

6) Self-actualized people have purpose.
"[They have] some mission in life, some task to fulfill, some problem outside themselves which enlists much of their energies. [...] This is not necessarily a task that they would prefer or choose for themselves; it may be a task that they feel is their responsibility, duty, or obligation. [...] In general these tasks are non personal or unselfish, concerned rather with the good of mankind in general."

> *Self-actualized people have the wonderful capacity to appreciate again and again, freshly and naïvely, the basic goods of life.*

7) They are not troubled by the small things. Instead, they focus on the bigger picture. "They seem never to get so close to the trees that they fail to see the forest. They work within a framework of values that are broad and not petty, universal and not local, and in terms of a century rather than the moment.[...] This impression of being above small things [...] seems to impart a certain serenity and lack of worry over immediate concerns that make life easier not only for themselves but for all who are associated with them."

8) Self-actualized people are grateful. They do not take their blessings for granted, and by doing so, maintain a fresh sense of wonder towards the universe. "Self-actualizing people have the wonderful capacity to appreciate again and again, freshly and naïvely, the basic goods of life, with awe, pleasure, wonder, and even ecstasy, however stale these experiences may have become to others [...] Thus for such a person, any sunset may be as beautiful as the first one, any flower may be of breath-taking loveliness, even after he has seen a million flowers. [...] For such people, even the casual workaday, moment-to-moment business of living can be thrilling."

> *Because of their self-decision, self-actualized people have codes of ethics that are individualized and autonomous.*

9) They share deep relationships with a few, but also feel identification and affection towards the entire human race.
"Self-actualizing people have deeper and more profound interpersonal relations than any other adults [...] They are capable of more fusion, greater love, more perfect identification, more obliteration of the ego boundaries than other people would consider possible. [...This devotion] exists side by side with a widespreading [...] benevolence, affection, and friendliness. These people tend to be kind [and

friendly] to almost everyone [...] of suitable character regardless of class, education, political belief, race, or color."

10) Self-actualized people are humble.
"They are all quite well aware of how little they know in comparison with what could be known and what is known by others. Because of this it is possible for them without pose to be honestly respectful and even humble before people who can teach them something."

11) Self-actualized people resist enculturation.
They do not allow themselves to be passively molded by culture — they deliberate and make their own decisions, selecting what they see as good, and rejecting what they see as bad. They neither accept all, like a sheep, nor reject all, like the average rebel. Self-actualized people: "make up their own minds, come to their own decisions, are self-starters, are responsible for themselves and their own destinies. [...] too many people do not make up their own minds, but have their minds made up for them by salesmen, advertisers, parents, propagandists, TV, newspapers and so on."

Because of their self-decision, self-actualized people have codes of ethics that are individualized and autonomous rather than being dictated by society. "They are the most ethical of people even though their ethics are not necessarily the same as those of the people around them [...because] the ordinary ethical behavior of the average person is largely conventional behavior rather than truly ethical behavior."

12) Despite all this, self-actualized people are not perfect.
"There are no perfect human beings! Persons can be found who are good, very good indeed, in fact, great. [...] And yet these very same people can at times be boring, irritating, petulant, selfish, angry, or depressed. To avoid disillusionment with human nature, we must first give up our illusions about it."

Because the brain enables behavior, to achieve the goals of Erikson and Maslow requires a non-compromised, healthy brain. In the pages that follow you will see how excessive stress, substance abuse, emotional and physical trauma, and more can increase the challenges for everyone. A brain not in optimal health diminishes the chances of achieving self actualization.

Sources:

Simplypsychology.com

NIH www.ncbi.nlm.nih.gov

Chapter Two

Exploring the Brain in Words and Pictures

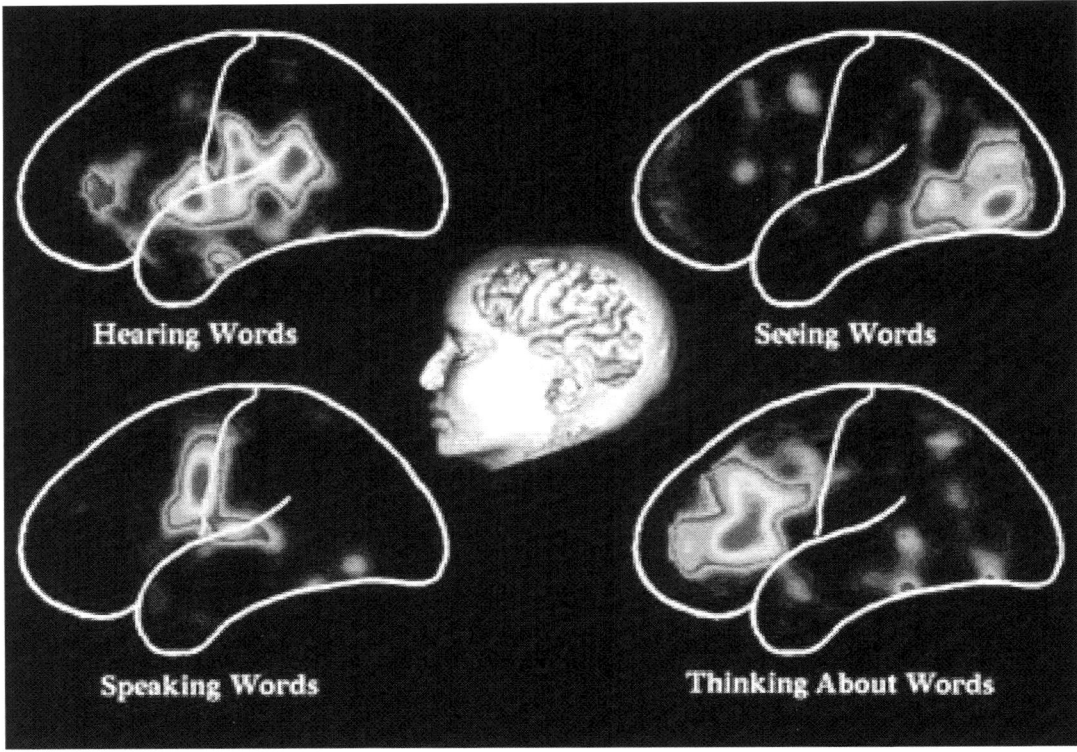

Dr. Marcus Raichle, Washington University, St. Louis, Department of Neuroscience. **These images demonstrate, by PET scanning (positron emission tomography), that certain areas of the brain activate as the brain performs specific language tasks.**

The brain has been called the most complex structure in the known universe. This three-pound pinkish-gray organ controls intelligence, interprets the senses, initiates body movement, and mediates behavior. Lying in its bony shell and washed by protective fluid, the brain is the source of all the qualities that define our humanity. The human brain represents the crown jewel of the human body, if not of all life.

For centuries, scientists and philosophers have been fascinated by it, but until recently they viewed the brain as nearly incomprehensible. Now, however, the brain is beginning to relinquish its secrets. Scientists have learned more about the brain in the last 30 years than in all previous centuries because of the accelerating pace of research in neurological and behavioral science and the development of new research techniques.

A Non-Stop Dynamic System
The brain is like a committee of experts. All the parts work together, but each part has its own special properties. The brain can be divided into three basic units: the forebrain, the midbrain, and the hindbrain.

The hindbrain includes the upper part of the spinal cord, the brain stem, and a wrinkled ball of tissue called the cerebellum. The hindbrain controls the body's vital functions such as respiration and heart rate. The cerebellum coordinates movement and is involved in learned rote movement.

The cerebrum sits at the topmost part of the brain and is the source of intellectual activities.

When you play the piano or hit a tennis ball, you are activating the cerebellum. The uppermost part of the brainstem is the midbrain, which controls some reflex actions and is part of the circuit involved in the control of eye movements and other voluntary movements. The forebrain is the largest and most highly developed part of the human brain: it consists primarily of the cerebrum and the structures hidden beneath it.

When people see pictures of the brain it is usually the cerebrum that they notice. The cerebrum sits at the topmost part of the brain and is the source of intellectual activities. It holds your memories, allows you to plan, enables you to imagine and think. It allows you to recognize friends, read books, and play games.

The cerebrum is split into two halves (hemispheres) by a deep fissure. Despite the split, the two cerebral hemispheres communicate with each other through a thick tract of nerve fibers that lies at the base of this fissure. Although the two hemispheres seem to be mirror images of each other, they are different. For instance, the ability to form words seems to lie primarily in the left hemisphere, while the right hemisphere seems to control many abstract reasoning skills.

For some as-yet-unknown reason, nearly all of the signals from the brain to the body and vice-versa cross-over on their way to and from the brain. This means that the right cerebral hemisphere primarily controls the left side of the body and the left hemisphere primarily controls the right side. When one side of the brain is damaged, the opposite side of the body is affected. For example, a stroke in the right hemisphere of the brain can leave the left arm and leg paralyzed.

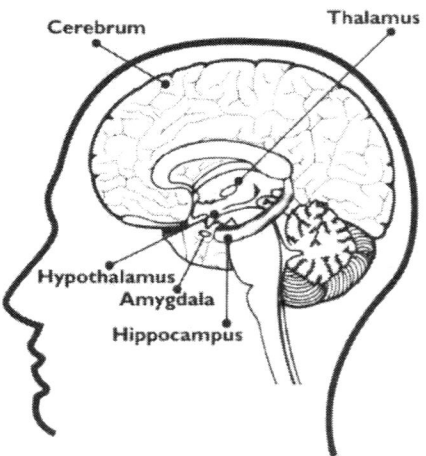

NIDA Drug Facts
Like a non-stop power plant that increased capacity while staying active, the brain has added higher-functioning areas above more primitive, yet essential, structures like the brain stem.

Where Thinking Happens

Each cerebral hemisphere can be divided into sections, or lobes, each of which specializes in different functions. To understand each lobe and its specialty we will take a tour of the cerebral hemispheres, starting with the two frontal lobes, which lie directly behind the forehead. When you plan a schedule, imagine the future, or use reasoned arguments, these two lobes do much of the work. One of the ways the frontal lobes seem to do these things is by acting as short-term storage sites, allowing one idea to be kept in mind while other ideas are considered. In the rearmost portion of each frontal lobe is a motor area, which helps control voluntary movement. A nearby place on the left frontal lobe called Broca's area allows thoughts to be transformed into words.

When you enjoy a good meal—the taste, aroma, and texture of the food—two sections behind the frontal lobes called the parietal lobes are at work. The forward parts of these lobes, just behind the motor areas, are the primary sensory areas. These areas receive information about temperature, taste, touch, and movement from the rest of the body. Reading and arithmetic are also functions in the repertoire of each parietal lobe.

As you look at the words and pictures on this page, two areas at the back of the brain are at work. These lobes, called the occipital lobes, process images from the eyes and link that information with images stored in memory. Damage to the occipital lobes can cause blindness.

The last lobes on our tour of the cerebral hemispheres are the temporal lobes, which lie in front of the visual areas and nest under the parietal and frontal lobes. Whether you appreciate symphonies or rock music, your brain responds through the activity of these lobes. At the top of each temporal lobe is an area responsible for receiving information from the ears. The underside of each temporal lobe plays a crucial role in forming and retrieving memories, including those associated with music. Other parts of this lobe seem to integrate memories and sensations of taste, sound, sight, and touch.

The Cerebral Cortex

Coating the surface of the cerebrum and the cerebellum is a vital layer of tissue the thickness of a stack of two or three dimes. It is called the cortex, from the Latin word for bark. Most of the actual information processing in the brain takes place in the cerebral cortex. When people talk about "gray matter" in the brain they are talking about this thin rind. The cortex is gray because nerves in this area lack the insulation that makes most other parts of the brain appear to be white. The folds in the brain add to its surface area and therefore increase the amount of gray matter and the quantity of information that can be processed.

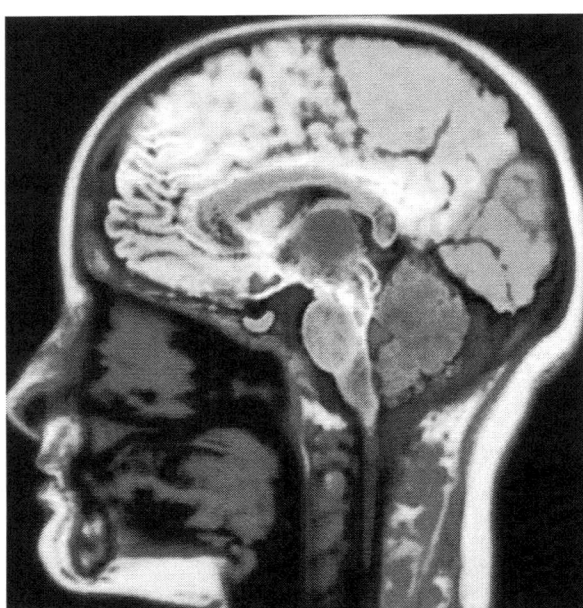

Alamy
Deep within the brain lie structures that are the gatekeepers between the spinal cord and the cerebral hemispheres.

The Inner Brain as Relay Station

Deep within the brain, hidden from view, lie structures that are the gatekeepers between the spinal cord and the cerebral hemispheres. These structures not only determine our emotional state, they also modify our perceptions and responses depending on that state, and allow us to initiate movements that you make without thinking about them. Like the lobes in the cerebral hemispheres, the structures described below come in pairs: each is duplicated in the opposite half of the brain.

The hypothalamus, about the size of a pearl, directs a multitude of important functions. It wakes you up in the morning, and gets the adrenaline flowing during a test or job interview. The hypothalamus is also an important emotional center, controlling the molecules that make you feel exhilarated, angry, or unhappy. Near the hypothalamus lies the thalamus, a major clearinghouse for information going to and from the spinal cord and the cerebrum.

An arching tract of nerve cells leads from the hypothalamus and the thalamus to the hippocampus. This tiny nub acts as a memory indexer—sending memories out to the appropriate part of the cerebral hemisphere for long-term storage and retrieving them when necessary.

The basal ganglia are clusters of nerve cells surrounding the thalamus. They are responsible for initiating and integrating movements. Parkinson's disease, which results in tremors, rigidity, and a stiff, shuffling walk, is a disease of nerve cells that lead into the basal ganglia.

All sensations, movements, thoughts, memories, and feelings are the result of signals that pass through neurons.

Making Connections
The brain and the rest of the nervous system are composed of many different types of cells, but the primary functional unit is a cell called the neuron. All sensations, movements, thoughts, memories, and feelings are the result of signals that pass through neurons. Neurons consist of three parts. The cell body contains the nucleus, where most of the molecules that the neuron needs to survive and function are manufactured. Dendrites extend out from the cell body like the branches of a tree and receive messages from other nerve cells.

Signals then pass from the dendrites through the cell body and may travel away from the cell body down an axon to another neuron, a muscle cell, or cells in some other organ. The neuron is usually surrounded by many support cells. Some types of cells wrap around the axon to form an insulating sheath. This sheath can include a fatty molecule called myelin, which provides insulation for the axon and helps nerve signals travel faster and farther.

Axons may be very short, such as those that carry signals from one cell in the cortex to another cell less than a hair's width away. Or axons may be very long, such as those that carry messages from the brain all the way down the spinal cord.

Neurotransmitters and the Relay Race of Life
Acetylcholine is called an *excitatory neurotransmitter* because it generally makes cells more excitable. It governs muscle contractions and causes glands to secrete hormones. Alzheimer's disease, which initially affects memory formation, is associated with a shortage of acetylcholine.

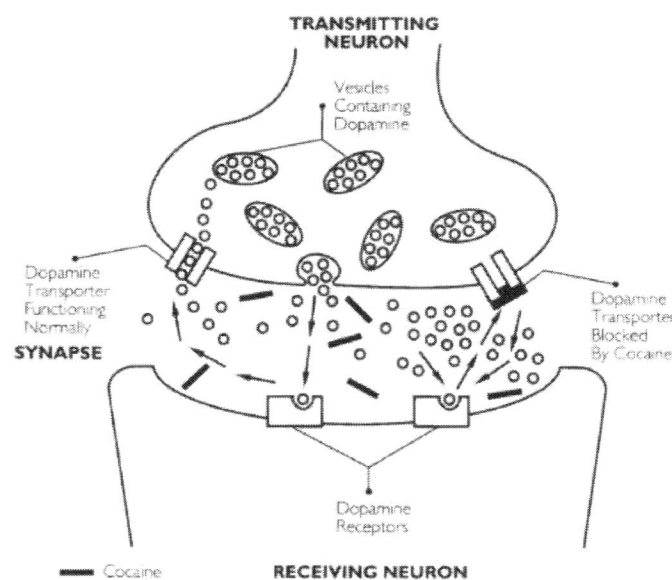

HHMI: Building a Genetic Map
Neurotransmitters communicate in the central nervous system from neuron to neuron via electro-chemical messaging across the synapse.

GABA (gamma-aminobutyric acid) is called an inhibitory neurotransmitter because it tends to make cells less excitable. It helps control muscle activity and is an important part of the visual system. Drugs that increase GABA levels in the brain are used to treat epileptic seizures and tremors in patients with Huntington's disease.

Diminished amounts of serotonin in spinal cords of suicide victims has been considered an important finding.

Serotonin, a workhorse neurotransmitter, is believed to play a key role in mood regulation, including states of depression. Diminished amounts of serotonin in spinal cords of suicide victims has been considered an important finding. We do not know if these findings apply to non-suicidal persons with depression. Serotonin constricts blood vessels and brings on sleep. It is also involved in temperature regulation.

Dopamine is an inhibitory neurotransmitter involved in mood and the control of complex movements. The loss of dopamine activity in some portions of the brain leads to the muscular rigidity of Parkinson's disease. Many medications used to treat behavioral disorders work by modifying the action of dopamine in the brain.

A Compromised Brain Can Bring Havoc

When the brain is healthy it functions quickly and automatically. But when problems occur, the results can be devastating. Some 50 million people in this country—one in five—suffer from damage to the nervous system. Some of the major types of disorders include: neurogenetic diseases (such as Huntington's disease and muscular dystrophy), developmental disorders (such as cerebral palsy), degenerative diseases of adult life (such as Parkinson's disease and Alzheimer's disease), metabolic diseases (such as Gaucher's disease), cerebrovascular diseases (such as stroke and vascular dementia), trauma (such as spinal cord and head injury), convulsive disorders (such as epilepsy), infectious diseases (such as AIDS dementia), and brain tumors.

> **With all the advances brain science has seen in the past decades, the future holds even greater promise.**

The Brain Imaging Revolution

Beginning in the early 1970s, advances in computer technology revolutionized brain research. Before then, the only people to have seen a living brain were brain surgeons. X-ray technology was unable to penetrate the skull. But with the aid of the computer, imaging technologies such as PET (positron emission tomography), CAT (computer assisted tomography), MRI (magnetic resonance imaging) and fMRI (functional magnetic resonance imaging) allowed surgeons and scientists to see the living brain in action.

Imaging techniques have revolutionized the field of brain science, enabling the precise mapping of brain functions and structures and permitting scientists to search out the roots of brain disorders or injuries. In addition, they have helped advance a "systems" view of brain function. According to this view, no one structure or area of the brain acts alone to drive a specific behavior or mental task. While certain brain areas may be specialized for certain tasks, brain function relies on networks of interconnected neurons. These specialized pathways enable the brain to analyze and assimilate information from external (e.g., sensory) as well as internal (e.g., hormones) cues in order to respond with appropriate physical and psychological behaviors. Systems neuroscience helps explain how people such as victims of stroke or head trauma, whose brains have been injured in a discrete site, can, over time, redevelop the functions lost as a result of the injury. Nerve cells in their brains in effect forge new pathways, bypassing the injured site and forming new connections, as if finding a new route to get to work after discovering that a bridge is out on the usual route.

Wikimedia.com
From childhood through the adult and aging years, the brain continually adapts, an ability called "plasticity."

This ability to adapt, which scientists call plasticity, seems to be particularly strong in young brains, but "old" brains routinely learn new tricks, scientists have found. Plasticity, in fact, plays a critical role in the entire life-cycle of the brain, from its development in infancy, to its continual reshaping as learning occurs, to its ability to adapt to age-related changes that can lead to mental deterioration in later life. Now, new evidence suggests the brain may be even more plastic than previously thought.

Turning one of the oldest tenets of neuroscience on its head, scientists recently discovered that nerve cells can regenerate, making the idea of brain repair following trauma or disease thinkable. Revealed at the end of the 20th century, this scientific breakthrough is sure to influence brain science for at least the next century.

Constructing the World's Most Sophisticated Machine
There is perhaps no time in the human life cycle during which plasticity is more important than in the period of nervous system development. A newborn baby's brain, scientists have learned, is not just a miniature version of an adult's. Instead, it is a work in progress, the world's most sophisticated machine in construction phase. Like the scaffolding that shapes the framework of a building, an initial framework of interneuronal "wiring" is present at birth, pre-set by nature via the genetic blueprints provided by the mother and father. The materials are also there: babies are born with virtually all of their lifetime store of nerve cells.

> *"Nurture" largely directs the completion of the wiring process, literally shaping the structure of the brain according to a child's early sensory experiences.*

What remains is the "finish work" of the brain's communications architecture, the fine-tuning of a quadrillion cell-to-cell connections. In humans, the fine-tuning phase unfolds over several developmental years. "Nurture" largely directs the completion of the wiring process, literally shaping the structure of the brain according to a child's early sensory experiences. During critical periods (or stages) of brain development, these early experiences stimulate neural activity in certain synaptic connections, which in turn become stronger and thrive. A "pruning" process ruled by a philosophy of "use it or lose it" ensues, during which synapses that are not routinely stimulated may wither and die. Within that period, "windows" of opportunity, during which the brain may be specially "primed" for learning certain skills such as language, open according to the developmental schedule of the brain regions underlying those skills. Since it's well known that humans can continue to learn and modify behavior throughout life, it's clear that the windows never really slam shut, even though they may become a bit sticky.

Children who fail to get the stimulation they need for proper brain development can become tragedies. In the 1990s, studies of Romanian orphans whose cries for comfort were never answered or whose smiles were never encouraged, found lingering impairments in the children's basic social and thinking abilities and in their physical development.

Numerous studies have also shown that babies who are held and caressed regularly do better developmentally and may reap the benefits throughout life. The first few years of life are especially important, as they are periods of rapid change in the synapses. But new understandings about the developing brain indicate that the process of fine-tuning connections among neurons continues, to varying degrees, into adolescence. In fact, "brain development" probably never really ends—older adults are also capable of forming new synaptic connections and do when they learn new things. But the rapid-paced period during which external stimuli are critical to "normal" brain-building generally begins to dwindle around the mid-teen years.

The brain's center for reason, advance planning, and other higher functions does not reach maturity until the early 20s.

Adolescence marks a turning point of sorts for the brain, as some of its structures are nearing maturity, while others are not yet fully developed. The prefrontal cortex, for example—the brain's center for reason, advance planning, and other higher functions—does not reach maturity until the early 20s. Since this part of the brain seems to act as a kind of cerebral "brake" to halt inappropriate or risky behaviors, some scientists believe sluggish development may explain difficulties in resisting impulsive behavior that some adolescents exhibit at times.

The brain also has ultimate control over the ebb and flow of powerful hormones such as adrenaline, testosterone, and estrogen, which themselves play critical roles in the changing adolescent body. The teenage brain is also struggling to adapt to a shift in the circadian rhythm, the brain's internal biological clock, which drives the sleep-wake cycle. The secretion of melatonin sets the timing for this internal clock, a hormone the brain produces in response to the daily onset of darkness. In one study, researchers found that the further along in puberty teens were, the later at night their melatonin was secreted. In practice, that means teens' natural biological clock is telling them to go to sleep later, and to stay asleep longer.

The Adult Brain: Attitude Counts!
While the teenage brain faces its share of challenges as it weathers the storm of adolescence, aging undoubtedly poses the greatest challenge to the normal life cycle of the brain. But contrary to popular belief, the slow march of mental decline many people associate with aging is not inevitable. While many people do experience memory lapses as they age, even as early as their 40s, this too is not preordained. Scientists who study the aging brain have identified an intriguing set of circumstances and personal attributes that seem to protect some people from the age-related declines in mental ability that so many aging Americans fear. In fact, maintaining mental sharpness into old age, attitude counts. Marilyn Albert, a neuropsychologist, and her colleagues have been follow a large group of "high-

Wikimedia Commons

Moderate to strenuous physical activity and higher levels of formal education have been found to be key predictors of cognitive maintenance.

her colleagues have been following a large group of "high-functioning" elderly people in an effort to determine what specific attributes tend to characterize people who maintain high levels of mental abilities into their 70s and beyond. Moderate to strenuous physical activity and higher levels of formal education have been found to be key predictors of cognitive maintenance. But perhaps the most surprising correlate with successful aging is a psychosocial factor that scientists call "self-efficacy."

Dr. Albert defines self-efficacy as "the feeling that what you do makes a difference in the things that happen to you every day. It boils down to feelings of control." Scientists have theorized that our self-efficacy beliefs influence the types of activities we pursue, as well as how much effort we put into them, and how persistent we are if the task proves difficult. If we have doubts about our ability to accomplish something, we may be less likely to try it, or may give up more easily. A cycle ensues: If we fail to engage in challenging activities, our risk for cognitive decline increases as we age. We might be anxious or stressed about what we can no longer do, which sets off a cascade of stress hormones that can themselves contribute to memory lapses, and may damage brain systems in other ways as well.

Scientists say that taking steps to assert control over one's life and surroundings, even in seemingly small ways, may help us to maintain our mental faculties well into old age. And, they say, it's never too early to begin.

What's a Quadrillion?

It's a number represented by 1 with 15 zeros (1,000,000,000,000,000). Scientists estimate that there are about 1 quadrillion synaptic connections between the neurons in the central nervous system, an estimate based on the belief that there are at least 100 billion (100,000,000,000) neurons, and that each neuron makes as many as 10,000 connections.

With numbers like these, it's no wonder we're just beginning to understand our brains. However, with all the advances brain science has seen in the past decades, the future holds even greater promise. By increasing our understanding of the brain, once considered a deep, unsolvable mystery, doctors, scientists, patients, and caregivers now have new hope for people suffering from some of our most devastating remaining diseases and disorders.

Wikimedia Commons
While an elephant's brain is physically larger than a human brain, the human brain is 2% of total body weight (compared to 0.15% of an elephant's brain).

A Brain Top Twenty

1. **No pain.** There are no pain receptors in the brain, so the brain can feel no pain.

2. **Largest brain.** While an elephant's brain is physically larger than a human brain, the human brain is 2% of total body weight (compared to 0.15% of an elephant's brain), meaning humans have the largest brain to body size.

3. **Stimulation**. A stimulating environment for a child can make the difference between a 25% greater ability to learn or 25% less in an environment with little stimulation.

4. **New neurons.** Humans continue to make new neurons throughout life in response to mental activity.

5. **Read aloud**. Reading aloud and talking often to a young child promotes brain development.

6. **10%.** The old adage of humans only using 10% of their brain is not true. Every part of the brain has a known function.

7. **Emotions.** The capacity for such emotions as joy, happiness, fear, and shyness are already developed at birth. The specific type of nurturing a child receives shapes how these emotions are developed,

8. **Child abuse and the brain.** Studies have shown that child abuse can inhibit development of the brain and can permanently affect brain development.

9. **Stress.** Excessive stress has shown to "alter brain cells, brain structure and brain function.

10. **Seafood.** In the March 2003 edition of *Discover* magazine, a report describes how people in a 7-year study who ate seafood at least one time every week had a 30% lower occurrence of dementia.

11. **Pain and gender**. Scientists have discovered that men and women's brains react differently to pain, which explains why they may perceive or discuss pain differently.

12. **Boredom.** Boredom is brought on by a lack of change or stimulation, is largely a function of perception, and is connected to the innate curiosity found in humans.

Wikimedia Commons
Coffee: America's stimulant of choice.

13. **Physical illness**. The connection between body and mind is a strong one. One estimate is that between 50-70% of visits to the doctor for physical ailments are attributed to psychological factors.

14. **Create associations**. Memory is formed by associations, so if you want help remembering things, create associations for yourself.

15. **Scent and memory**. Memories triggered by scent have a stronger emotional connection, therefore appear more intense than other memory triggers.

16. **Everyone dreams**. Just because you don't remember your dreams doesn't mean you don't dream.

17. **Caffeine.** Caffeine works to block naturally occurring adenosine in the body, creating alertness. Scientists have recently discovered this connection and learned that doing the opposite—boosting adenosine—can actually help promote more natural sleep patterns and help eliminate insomnia.

Pixabay.com

By memorizing the city's complicated street patterns, London taxi drivers have developed an unusually strong type of memory.

18. **Music**. Music lessons have shown to considerably boost brain organization and ability in both children and adults.

19. **Albert Einstein**. Einstein's brain was similar in size to other humans except in the region that is responsible for math and spatial perception. In that region, his brain was 35% wider than average.

20. **London taxi drivers**. Famous for knowing all the complicated London streets by heart, these drivers have a larger than normal hippocampus, especially the drivers who have been on the job longest. The study suggests that as people memorize more and more information, this part of their brain continues to grow.

Sources

NIH.gov

NINDS.gov

Dana Sourcebook of Brain Science, Third Edition

Chapter Three

Use It or Lose It
Maintaining Brain Health

By David Mahoney and Richard M. Restak, M.D.

(Book excerpt.) Authors David Mahoney, businessman and philanthropist, and neurologist Richard Restak, M.D., point out that the 20th century's increase in life span (30 years longer average life expectancy) and spectacular gains in neuroscience and medicine will make the 100-year life span commonplace in another generation o two. This advance will happen only if we take proactive measures in such matters as handling stress properly and seeking out lifelong mental activity.

Pixabay
Neglecting your tennis or golf for enough time and your skills in these very different activities will deteriorate.

Certain cells in areas of the brain beneath the cortex (called subcortical nuclei) are sometimes irreverently dubbed the "juice machines." They give us enthusiasm and general "get up and go" energy. When Samuel Johnson said, "The question is not so much 'Is it worth seeing?' but rather 'Is it worth going to see?'" he was unknowingly referring to the subcortical nuclei, which generate enthusiasm and energy.

With aging, almost everyone undergoes some loss of cells in the subcortical nuclei. It's what we notice when we joke that our get-up-and-go "got up and went." Since this is natural, our task is to recognize that we are "mellowing" rather than losing any of our abilities. Our attention to those abilities, as intact as ever, helps us maintain mental vigor. Every talent and special skill that you've developed over your lifetime is represented in your brain by a complex network of neurons. And each time you engage in any activity that involves your talents and skills, the neuronal linkages in that network are enhanced. Think of the brain cells as shaped like trees composed of long branches subdividing into smaller and smaller branches.

As the result of brain growth and the person's experience in the world, tremendous overlap and connectivity develop among the tree branches. Neuroscientists, struck with the tree analogy, refer to this process as "arborization."

> ***No matter how long you've neglected a skill, you'll never find yourself in the same situation as the person who never learned the skill in the first place.***

Eventually nerve cells form active circuits based on these branchlike linkages. The more often the circuits are activated, the easier it is to activate them the next time. Subjectively, you experience this as the formation of a habit. With time the activity gets easier to do; the more the skill or talent is practiced, the better you get at it. But if you neglect your talents and skills, they begin to wane, and over time it becomes harder and harder to perform at your best. If enough time passes you will experience great difficulty returning to your former level of excellence. That's because the neuronal circuits have fallen into disuse: greater degrees of effort are required to activate them.

But no matter how long you've neglected a skill, you'll never find yourself in the same situation as the person who never learned the skill in the first place. Neuronal circuits, once established, never entirely disappear. It's the ease of facilitating them that varies. This law of facilitation and disuse atrophy applies to every activity, whether physical or mental.

> **When we stop challenging ourselves and expanding, or at least maintaining our skills, the brain cells involved in the neuronal networks drop out and link into other networks.**

Neglect your tennis or your golf for enough time and your skills in these very different activities will deteriorate. Remember that the brain is an ever-changing organ. If one part gets rusty and suffers atrophy from disuse, its functions are taken over by other areas that are used more. When we stop challenging ourselves and expanding, or at least maintaining our skills, the brain cells involved in the neuronal networks drop out and link into other networks. Eventually the skill has almost entirely disappeared. We say almost because some neurons, though a much smaller number, always remain in the network.

(Excerpted from The Longevity Strategy: How to Live to 100 Using the Brain-Body Connection, *by David Mahoney and Richard Restak, M.D. John Wiley & Sons, Inc., and Dana Press, New York, 1998.)*

Chapter Four

The Power of Emotions

By Joseph E. LeDoux, Ph.D.

(Book excerpt.) New York University neuroscientist Joseph LeDoux, Ph.D., and other neuroscientists have begun to examine the way the brain shapes our experience—and our memories—to generate the varied repertoire of human emotions. Specifically, as Dr. LeDoux explains, he chose to begin his inquiry by examining an emotion that is common to all living creatures: fear.

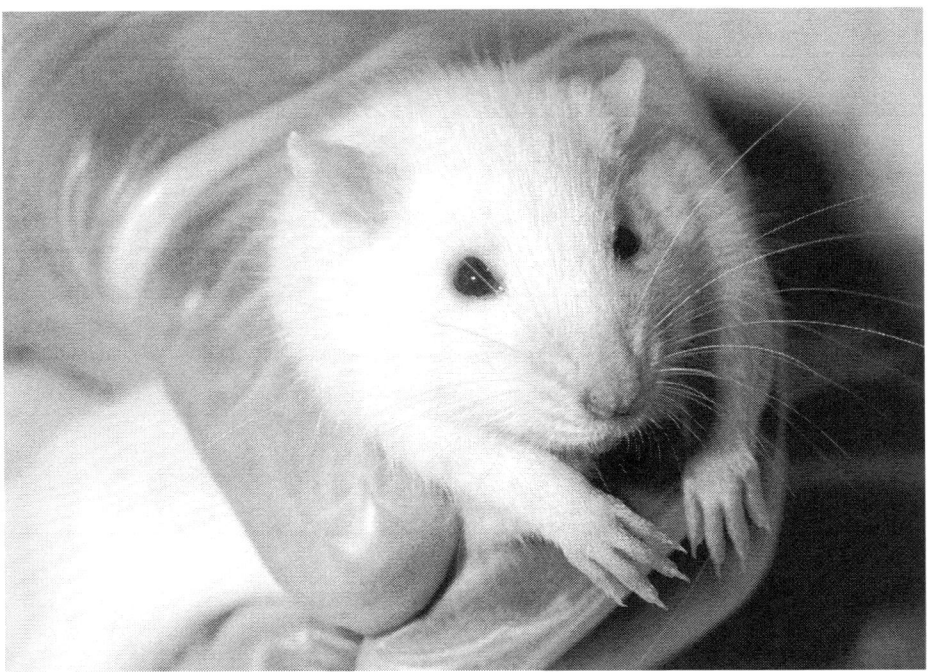

Wikimedia.com
Mice serve researchers well as animal models. These very distant relatives possess well over 90 per cent of the same genes as humans.

Years of research by many workers have given us extensive knowledge of the neural pathways involved in processing acoustic information, which is an excellent starting point for examining the neurological foundations of fear. The natural flow of auditory information—the way you hear music, speech, or anything else—is that the sound comes into the ear, enters the brain, goes up to a region called the auditory midbrain, then to the auditory thalamus, and ultimately to the auditory cortex.

Thus, in the auditory pathway, as in other sensory systems, the cortex is the highest level of processing.

So the first question we asked when we began these studies of the fear system was: Does the sound have to go all the way to the auditory cortex in order for the rat to learn that the sound paired with the shock is dangerous? When we made lesions in the auditory cortex, we found that the animal could still make the association between the sound and the shock, and would still react with fear behavior to the sound alone. Since information from all our senses is processed in the cortex—which ultimately allows us to become conscious of seeing the predator or hearing the sound—the fact that the cortex didn't seem to be necessary to fear conditioning was both intriguing and mystifying. We wanted to understand how something as important as the emotion of fear could be mediated by the brain if it wasn't going into the cortex, where all the higher processes occur.

> ***Some other area or areas of the brain must receive information from the thalamus and establish memories about experiences that stimulate a fear response.***

So we next made lesions in the auditory thalamus and then in the auditory midbrain. The midbrain supplies the major sensory input to the thalamus, which in turn supplies the major sensory input to the cortex. What we found was that lesions in either of these subcortical areas completely eliminated the rat's susceptibility to fear conditioning. If the lesions were made in an unconditioned rat, the animal could not learn to make the association between sound and shock, and if the lesions were made on a rat that had already been conditioned to fear the sound, it would no longer react to the sound. But if the stimulus didn't have to reach the cortex, where was it going from the thalamus?

Some other area or areas of the brain must receive information from the thalamus and establish memories about experiences that stimulate a fear response. To find out, we made a tracer injection in the auditory thalamus (the part of the thalamus that processes sounds) and found that some cells in this structure projected axons into the amygdala. This is key, because the amygdala has for many years been known to be important in emotional responses. So it appeared that information went to the amygdala from the thalamus without going to the neocortex. We then did experiments with rats that had amygdala lesions, measuring freezing and blood pressure responses elicited by the sound after conditioning. We found that the amygdala lesion prevented conditioning from taking place. In fact, the responses are very similar to those of unconditioned animals that hear the sound for the first time, without getting the shock. So the amygdala is critical to this pathway.

It receives information about the outside world directly from the thalamus, and immediately sets in motion a variety of bodily responses. We call this thalamo-amygdala pathway the low road because it's not taking advantage of all of the higher-level information processing that occurs in the neocortex, which also communicates with the amygdala.

(Excerpted from States of Mind: New Discoveries About How Our Brains Make Us Who We Are, *Roberta Conlan, editor. Dana Press and John Wiley & Sons, Inc., New York, 1999.)*

Chapter Five

Wounds that Time Alone Won't Heal
The Biology of Stress

Imagine you are a zebra grazing on the plains of Africa. It's midday. The sun is bright, the food is plentiful.

Suddenly you sense an attack. A lion is chasing you. Its fight or flight in action.

Your brain tells your body to prepare for a fight or take flight. The body responds by preparing extra hormones to create more energy and by increasing the rate the heart pumps blood to the muscles. For most animals, this stress reaction lasts for just a short time and it saves lives.

Wikimedia Commons
Why don't zebras get ulcers? According to Dr. Robert Sapolsky, their stress is decidedly short term, not long term.

As a body is preparing for fight or flight, however, practically all systems, such as digestion, physical growth, and warding off diseases are placed on hold. This means that people for whom stress has become a way of life are endangering their overall health. Researchers have learned by studying primates whose systems are similar to human beings that those who learn to have control over their lives and are able to reduce or avoid stress live longer and healthier lives.

Are zebras better equipped to deal with stress than humans? No. However, according to Dr. Robert Sapolsky, author of *Why Zebras Don't Get Ulcers*, "For a zebra, stress is three minutes of some screaming terror running from a lion. After the chase, either it's over or they are." On the other hand humans, he says, have constructed a network of social stressors. Since we are obliged to live in this framework, stress builds up.

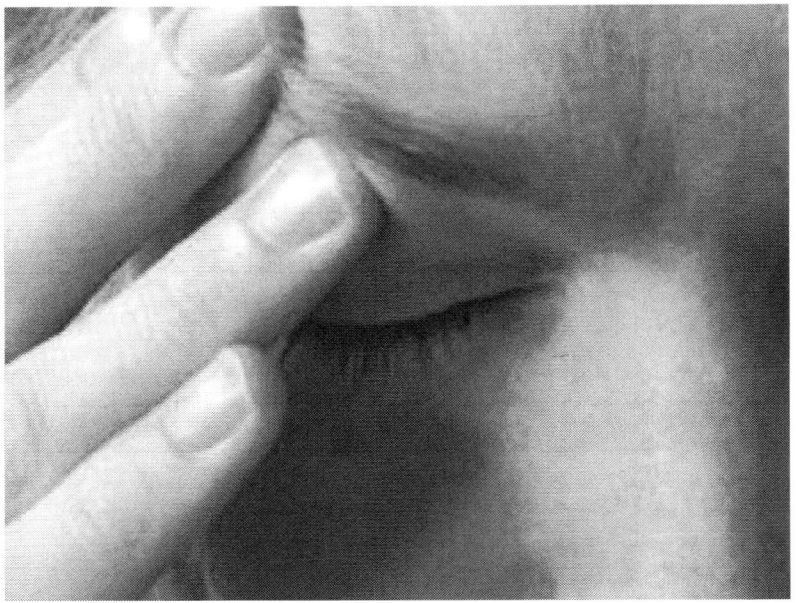

Nature.com
While the stress response activates automatically, its duration and intensity relies on factors such as individual temperament.

How do the brain and the body react to stress? Stress, such as the threat of attack, forces various changes in the body. First, adrenaline causes an increase in heart rate and blood pressure so that blood can be sent to muscles faster. Second, the brain's hypothalamus signals the pituitary gland to stimulate the adrenal gland (specifically the adrenal cortex) to produce cortisol.

This stress hormone, a longer-acting steroid, helps the body to mobilize energy. However, prolonged exposure to cortisol can damage virtually every part of the body. Chronic high blood pressure can cause blood vessel damage and the long-term shutdown of digestion can lead to ulcers.

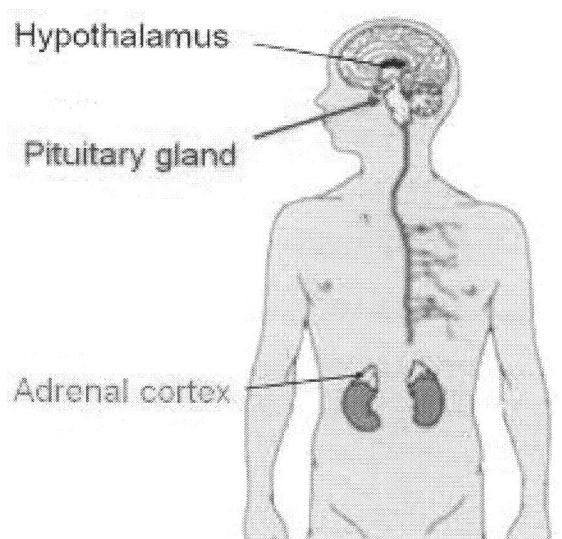

SimplyPsychology.org
Stress, such as the threat of attack, forces changes in the body carried out by the hypothalamus-pituitary-adrenal axis (HPA).

Why do some people experience more stress than others? Individuals who feel they have control over their lives appear to experience less stress. It also depends on personality and temperament. Aggressive, competitive types are more likely to define a situation as stressful than a passive, accommodating personality. A universal stress producer seems to be social isolation.

PTSD: A Breakthrough in Diagnosis

In 1980 the mental health community established the diagnosis of Post Traumatic Stress Disorder, PTSD, and revolutionized the way the field views the effects of stress. This change acknowledged that many of the symptoms people experience after exposure to trauma can be long-lasting, if not permanent. Before that shift, the field tended to view stress-related symptoms as a transient, normal response to an adverse life event, not requiring intensive treatment.

Furthermore, before 1980, people who did develop long-term symptoms following trauma were viewed as implicitly vulnerable; the role of the actual event in precipitating their symptoms was minimized. For a while, in a reversal of previous thinking, experts expected most trauma survivors to develop PTSD. More recent research has confirmed that only about 25 per cent of individuals who are exposed to trauma develop PTSD.

So who is likely to develop PTSD following a traumatic experience, and why? The answer is not yet clear, but it now appears that PTSD represents a failure of the body to extinguish or contain the normal nervous system response to stress. This failure is associated with many factors:

- **the nature and** severity of the traumatic event
- **preexisting risk factors** related to previous exposure to stress or trauma, particularly in childhood
- **the individual's history** of psychological and behavioral problems, if any
- **the person's level** of education, and other cognitive factors
- **family history—whether** parents or other relatives had anxiety, depression, or PTSD

People who develop PTSD are also more likely to develop other psychiatric disorders involving mood (depression, anxiety and panic, bipolar disorder), personality, eating, and substance dependence.

People also seek medical help for problems that may develop after the trauma that can mask or intensify PTSD symptoms. These symptoms include chronic pain, fatigue, headaches, muscle cramps, and self-destructive behavior, including alcohol or drug abuse and suicidal gestures. Often, survivors are not aware that their physical symptoms are related to their traumatic experiences. They may even fail to mention those disturbing events to their physicians, which can make PTSD difficult to diagnose accurately.

The Vulnerable Brains of Children and Youth
Our brains are shaped by our early experiences. Poor treatment is a chisel that shapes a brain to contend with conflict, but at the cost of deep, enduring wounds. Childhood abuse isn't something you "get over." It is a danger we must acknowledge and confront if we aim to do anything about the cycle of violence in this country.

One can easily understand how beating a child may damage the developing brain, but what about the all-too-common psychological abuse of children? Because the abuse was not physical, these children may be told, as adults, that they should just "get over it."

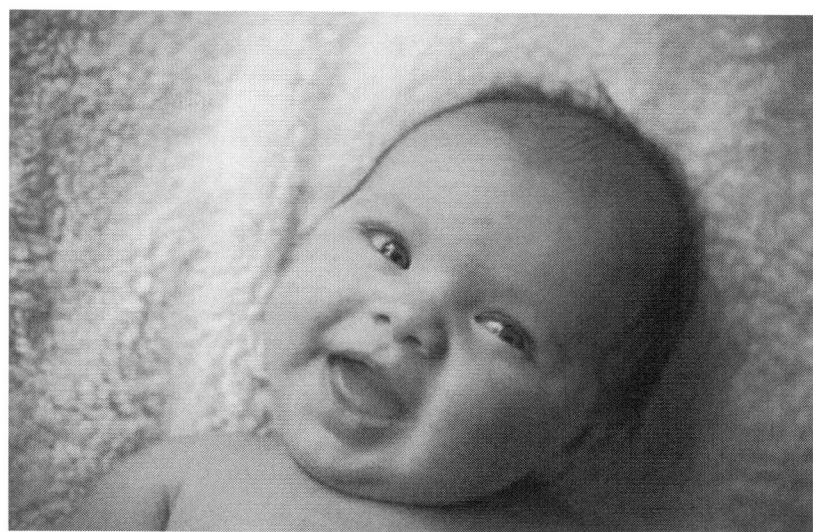

FreeBeacon.com
Early experiences guide the development of our brains. Poor treatment is a chisel that shapes a brain to contend with conflict.

Their Choice or Ours?
Evidence is accumulating that trauma, especially early in life, repeated, and inflicted by relatives or caretakers, produces dissociative disorders. These disorders can be thought of as a chronic, severe form of post-traumatic stress disorder. The essence of traumatic stress is helplessness—a loss of control over one's body. The mental imprint of such frightening experiences sometimes takes the form of loss of control over parts of one's mind—identity, memory, and consciousness—just as physical control is regained.

> ***Victims have reported being dazed, unaware of serious physical injury, or experiencing the trauma as if they were in a dream.***

During and in the immediate aftermath of acute trauma, such as an automobile accident or a physical assault, victims have reported being dazed, unaware of serious physical injury, or experiencing the trauma as if they were in a dream.

Many rape victims report floating above their body, feeling sorry for the person being assaulted below them. Sexually or physically abused children often report seeking comfort from imaginary playmates or imagined protectors, or by imagining themselves absorbed in the pattern of the wallpaper. Some continue to feel detached and disintegrated for weeks, months or years after trauma.

PTSD in Children and Teens
Children and teens could have PTSD if they have lived through an event that could have caused them or someone else to be killed or badly hurt. Such events include sexual or physical abuse or other violent crimes. Disasters such as floods, school shootings, car crashes, or fires might also cause PTSD. Other events that can cause PTSD are war, a friend's suicide, or seeing violence in the area they live.

Also, three to ten million children witness family violence each year. Around 40 to 60 per cent of those cases involve child physical abuse.

Children and teens who go through the most severe traumas tend to have the highest levels of PTSD symptoms. The PTSD symptoms may be less severe if the child has more family support and if the parents are less upset by the trauma. Lastly, children and teens who are farther away from the event report less distress.

Other factors can affect PTSD. Events that involve people hurting other people, such as rape and assault, are more likely to result in PTSD than other types of traumas. Also, the more traumas a child goes through, the higher the risk of getting PTSD.

Harsh Choices for Children and Youth
Abuse by a trusted authority figure such as a parent creates special problems. A child abused by a family member faces an ongoing dilemma: this beloved figure is inflicting harm, pain, and humiliation, yet the child is both emotionally and physically dependent. The child has to maintain two opposing views of the same person, which creates considerable tension and confusion, a situation described by one psychologist as "betrayal trauma." This psychologist showed that people prone to dissociation have selective amnesia for trauma-related words such as "incest."

What does PTSD look like in infants and children?
Animal models have taught us that stressing the mother in pregnancy can alter brain development in the offspring; and that prolonged separation of infant from mother impairs in the newborn other aspects of brain development and function. Furthermore, inconsistent maternal care and maternal anxiety, for example, from food insecurity, produce anxiety in offspring and contribute to the predisposition to diabetes, which itself has adverse effects on the brain.

In one of the most striking examples, an infant rat becomes attracted to odors from the mother early in life before the fear system is developed, and this attraction can occur even when the infant is abused. This paradoxical attraction of infant to the abusing mother allows the pup to survive, because mother is the only source of nutrition. Indeed, the presence of the mother suppresses development of a brain structure involved in fear and aversive learning. Translated into human terms, this phenomenon may help explain the behavior of individuals who are abused and neglected as children and yet who may choose a partner similar to an abusing parent.

Studies on children growing up in adversity have added to the information gained from animal research. Chaos in the home and inconsistent parenting impairs development of self regulatory behaviors, which can lead to substance abuse, earlier onset of sexual activity, bad decision making, and poor mood control.

> **Some PTSD symptoms in teens may begin to look like those of adults. One difference is that teens are more likely than younger children or adults to show impulsive and aggressive behaviors.**

School-aged children (ages 5-12)
These children may not have flashbacks or problems remembering parts of the trauma, the way adults with PTSD often do. Children, though, might put the events of the trauma in the wrong order. They might also think there were signs that the trauma was going to happen. As a result, they think that they will see these signs again before another trauma happens. They think that if they pay attention, they can avoid future traumas.

Children of this age might also show signs of PTSD in their play. They might keep repeating a part of the trauma. These games do not make their worry and distress go away. For example, a child might always want to play shooting games after he sees a school shooting. Children may also fit parts of the trauma into their daily lives. For example, a child might carry a gun to school after seeing a school shooting.

Teens (ages 12-18)
Some PTSD symptoms in teens may begin to look like those of adults. One difference is that teens are more likely than younger children or adults to show impulsive and aggressive behaviors.

Wikimedia Commons
Teen experience of PTSD can mimic those of adults and include re-experiencing a traumatic event, flashbacks, and nightmares.

Adult symptoms may include:
- Re-experiencing the traumatic event
- Increased anxiety and emotional arousal
- Intrusive, upsetting memories of the event
- Flashbacks (acting or feeling like the event is happening again)
- Nightmares (either of the event or of other frightening things)
- Feelings of intense distress when reminded of the trauma
- Loss of interest in activities and life in general
- Sense of a limited future (not expecting to live a normal lifespan, get married, have a career)

Below the surface, some children from deprived surroundings, have vastly different hormone levels than their parent-raised peers even beyond the baby years. Studies have shown that children who experienced early deprivation also had different levels of oxytocin and vasopressin (hormones that have been linked to emotion and social bonding), despite having had an average of three years in a family home.

It has been thought that these changes in hormones and neurotransmitters impair development of vulnerable brain regions. If we observe an association between a history of abuse and the presence of a physical abnormality, the abuse may have caused that abnormality. But it is also possible that the abnormality occurred first and elevated the likelihood of abuse, or that the abnormality ran in the family and led to more frequent abusive behavior by family members or other relatives.

> ***People with PTSD actively avoid situations that might bring back memories of the trauma.***

A Harvest of Psychiatric Disorders

Changes to normal body chemistry induced by physical, sexual, and psychological trauma in childhood may lead to psychiatric difficulties that show up in childhood, adolescence, or adulthood. The victim's anger, shame, and despair can be directed inward to produce symptoms such as depression, anxiety, and suicidal ideation, or directed outward as aggression, impulsiveness, delinquency, hyperactivity, and substance abuse.

Childhood trauma may fuel a range of persistent psychiatric disorders. One is somatoform disorder (also known as psychosomatic disorder), in which patients experience physical complaints with no discernible medical cause. Another is panic disorder with agoraphobia (fear of open spaces) in which patients experience the sudden, acute onset of terror and may narrow their range of activities to avoid being outside, especially in public, in case they have an attack.

People with PTSD keep re-experiencing a traumatic event in waking life or in dreams, and they actively avoid situations that might bring back memories of the trauma.

More complex, difficult-to-treat disorders strongly associated with childhood abuse are borderline personality disorder and dissociative identity disorder. Someone with borderline personality disorder characteristically sees others in black-and-white terms, first putting them on a pedestal, then vilifying them after some perceived slight or betrayal. Such people have a history of intense but unstable relationships, feel empty or unsure of their identity, often try to escape through substance abuse, and experience self-destructive impulses and suicidal thoughts. They are plagued by anger, most often directed at themselves.

In dissociative identity disorder, formerly called multiple personality disorder (the phenomenon behind Robert Louis Stevenson's "Dr. Jekyll and Mr. Hyde"), at least two seemingly separate people occupy the same body at different times, each with no knowledge of the other. This can be seen as a more severe form of borderline personality disorder. In borderline personality disorder, there is one dramatically changeable personality with an intact memory, as opposed to several distinct personalities, each with an incomplete memory.

Of the many disorders associated with childhood abuse, depression or heightened risk for developing it, may be a consequence of reduced activity of the left frontal lobes. If so, the stunted development of the left hemisphere related to abuse could easily enhance the risk of developing depression. Similarly, excess electrical irritability in the limbic system, and alterations in development of receptors that modulate anxiety, set the stage for the emergence of panic disorder and increase the risk of post-traumatic stress disorder.

> ***Thirty per cent of children with a history of severe abuse meet the diagnostic criteria for attention-deficit / hyperactivity disorder (ADHD)***

Alterations in the neurochemistry of these areas of the brain also heighten the hormonal response to stress, producing a state of hypervigilance and right-hemisphere activation that colors our view with negativity and suspicion. Alterations in the size of the hippocampus, along with limbic abnormalities shown on an EEG, further enhance the risk for developing dissociative symptoms and memory impairments.

Researchers have also found that 30 per cent of children with a history of severe abuse meet the diagnostic criteria for attention-deficit/hyperactivity disorder (ADHD), although they are less hyperactive than children with classic ADHD. Very early childhood abuse appears particularly likely to be associated with emergence of ADHD-like behavior problems. Interestingly, one of the most reliable brain structure findings in ADHD is reduced size of the cerebellar vermis. The cerebellar vermis receives information from the spinal cord about the sense of touch and proprioception. Proprioception is the ability to sense or perceive the spatial position and movements of the body. The cerebellar vermis also receives information from the body about hearing, vision, and balance.

Some studies have also found an association between reduced size of the mid portions of the corpus callosum and emergence of ADHD-like symptoms of impulsivity. Hence, early abuse may produce brain changes that mimic key aspects of ADHD.

An Increased Startle Response?
Researchers also think that childhood trauma may lead to what is called an exaggerated "startle response" on into adulthood. When startled, people experience a number of reactions. The heart may race, sweat increase, breath rate get faster, muscles tense (to the point someone might even jump), and people may feel scared. When someone jumps out from behind and yells, "Boo!" that may initiate a startle response. This is an ongoing area of investigation.

Wikimedia Commons
Policies of a repressive regime scarred a generation of Romanian youth.

Traumatized Children and Youth in Romania--A Tragedy of Epic Proportions

Beginning in the 1960s, the country of Romania's harsh economic policies meant that most families were too poor to support multiple children. So, without other options, thousands of parents left their babies in government-run orphanages.

By Christmas day 1989, when revolutionaries overthrew the government, an estimated 170,000 children were living in more than 700 state orphanages. As the regime crumbled, journalists and humanitarians swept in. In most institutions, children were getting adequate food, hygiene and medical care, but had woefully few interactions with adults, leading to severe behavioral and emotional problems.

Unlike growing up in a family, the children didn't have lots of interactions with adults holding them, talking to them, singing or playing with them, and that lack of stimulation affected their brain development.

An American scientist who went to study the crisis, recalls "a boy in a red T-shirt and sweats skipped up to me, grabbed my hand, and wouldn't let go. His head didn't reach my shoulders, so I figured he was eight or nine years old. He was 13, my guide said. The boy kept looking up at me with an open, sweet face, but I found it difficult to return his gaze.

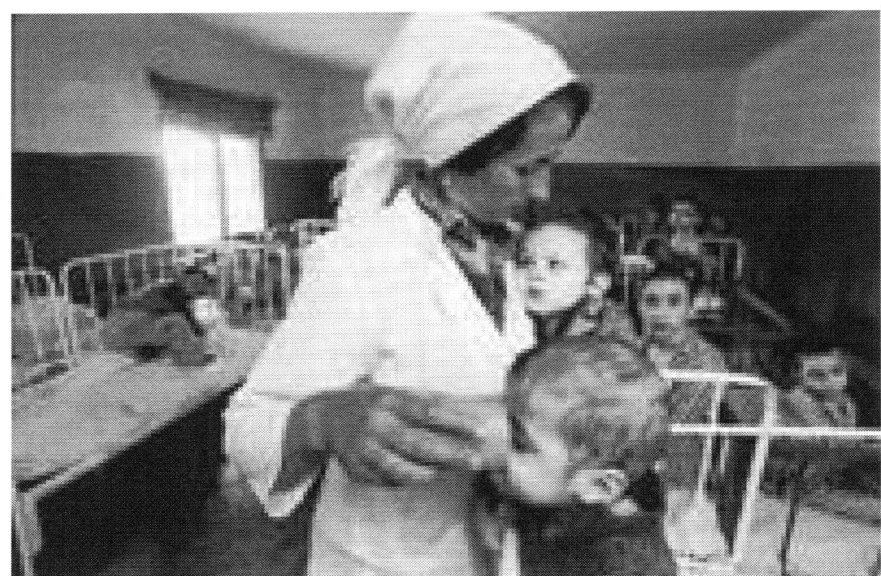

Wikimedia Commons
Harsh economic conditions and government actions beginning in the 1960s forced many families in Romania to abandon their children to state homes.

Like most of the other kids, he had crossed eyes — strabismus, a professor would explain later, a common symptom of children raised in institutions, possibly because as infants they had nothing to focus their eyes on. A couple of dozen kids gathered around us in a tight circle, chirping and giggling loudly as children do. At one point they broke into a laughing fit, and I asked my guide what happened. They were gawking at the whiteness of my teeth, he said. Two of the girls, somewhere in that gaggle, were pregnant.

Children were getting adequate food, hygiene and medical care, but had woefully few interactions with adults, leading to severe behavioral and emotional problems

> ***Mary spoon-fed him a mashed banana. He reacted with surprise. "It was very odd and strange to him."***

Mary Barrett, a prospective American adoptive mother recalls that she met an 11-month-old named Daniel. "He had an eagerness," she remembers. "He was alert. He was cruising the side of his crib and looking for stimulation." The small boy had been in a hospital crib his whole life and fed only by bottle. Mary spoon-fed him a mashed banana. He reacted with surprise she recalls, "It was very odd and strange to him."

The Barretts adopted Daniel when he was 13 months old. He was small for his age, scoring in the fifth percentile of height and weight. They thought it would be a matter of "playing catch-up," says Mary. That it was "a case of delay that would be overcome by paying extra attention." She says she remained optimistic for two years. But certain things didn't seem right.

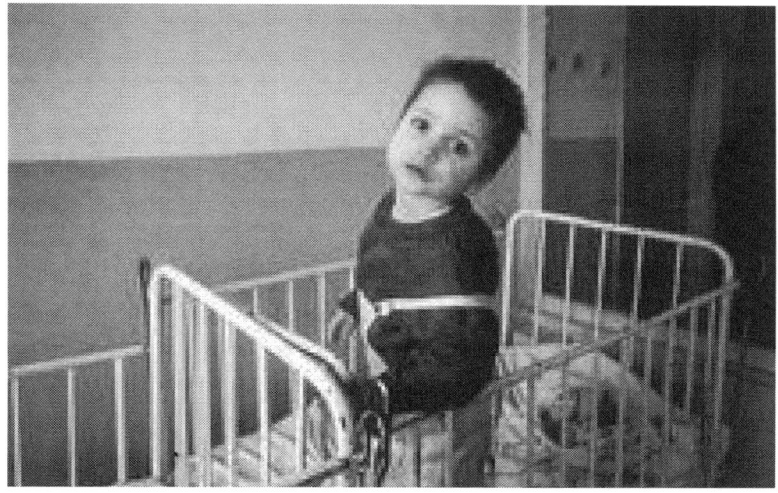

Wikimedia Commons
At state-run institutions in Romania, children received food, hygiene, and medical care, but had woefully few interactions with adults, leading to severe physical behavioral and emotional distress.

"We'd call 'Daniel' and usually a child responds to his name," she says. "[It's] one of the first things." Daniel didn't seem interested in the sound of a person's voice. The hearing tests came back normal. Daniel had a more profound communication problem. "He didn't realize people were the center of the universe," says his mother.

The pre-schooler was much more interested in the mechanical world, a characteristic of children with autism. "People were less important than a leaf flying in the wind," she remembers.

Wikimedia Commons
Infants and toddlers require skin-to-skin contact for proper emotional and physical development.

How Is PTSD Treated in Children and Teens?
The Healing Power of Touch

We have known for a long time that skin-to-skin contact with babies is important for their development. In what ways does it help them? Touch at stages of life, but particularly for infants and children, is life-saving. Its absence can be devastating.

According to anthropologist Robin Dunbar, touch works because it stimulates the release of endorphins--natural opiates produced by the brain that trigger feelings of relaxation by lowering the heart rate, reducing overt nervous behaviors, and even bringing on sleep.

Particularly in the newborn period, touch helps calm babies: They cry less and sleep more soundly. There are some studies that show their brain development is enhanced—probably because babies are calmer and more relaxed.

Does skin-to-skin contact with their babies have benefits for the parents? It seems to help the mothers, too. It reduces their stress level—they report lower levels of depression, they seem to be able to be more sensitive to their baby's cues and the babies are more responsive to the mother through the whole first three months. They're recognizing their mother earlier, so the relationship between the mother and baby is off to a good start. It works the same way with fathers, too.

For many children, PTSD symptoms go away on their own after a few months. Yet some children show symptoms for years if they do not get treatment.

Treatment Options for PTSD in Children
Cognitive-Behavioral Therapy (CBT)
CBT has been called the most effective approach for treating children. One type of CBT is called Trauma-Focused CBT (TF-CBT). In TF-CBT, the child may talk about his or her memory of the trauma. TF-CBT also includes techniques to help lower worry and stress. The child may learn how to assert himself or herself. The therapy may involve learning to change thoughts or beliefs about the trauma that are not correct or true. For example, after a trauma, a child may start thinking, "the world is totally unsafe."

Some may question whether children should be asked to think about and remember events that scared them. However, this type of treatment approach is useful when children are distressed by memories of the trauma. The child can be taught at his or her own pace to relax while they are thinking about the trauma. That way, they learn that they do not have to be afraid of their memories.

CBT often uses training for parents and caregivers as well. It is important for caregivers to understand the effects of PTSD. Parents need to learn coping skills that will help them help their children.

Psychological First Aid/Crisis Management
Psychological First Aid (PFA) has been used with school-aged children and teens that have been through violence where they live. PFA can be used in schools and traditional settings. It involves providing comfort and support, and letting children know their reactions are normal. PFA teaches calming and problem solving skills. PFA also helps caregivers deal with changes in the child's feelings and behavior. Children with more severe symptoms may be referred for added treatment.

Play Therapy
Play therapy can be used to treat young children with PTSD who are not able to deal with the trauma more directly. The therapist uses games, drawings, and other methods to help children process their traumatic memories.

Special treatments may be needed for children who show out-of-place sexual behaviors, extreme behavior problems, or problems with drugs or alcohol abuse.

Strong or Traumatized Youth--We Reap What We Sow
Whether abuse of a child is physical, psychological, or sexual, it sets off a ripple of hormonal changes that wire the child's brain to cope with a hostile world. Abuse predisposes the child to have a biological basis for fear, though he may act and pretend otherwise. Early abuse molds the brain to be more irritable, impulsive, suspicious, and prone to be swamped by fight-or-flight reactions that the rational

mind may be unable to control. The brain is programmed to a state of defensive adaptation, enhancing survival in a world of constant danger, but at a terrible price.

To a brain so tuned, the Garden of Eden would seem to hold its share of dangers; building secure, stable relationships may later require extraordinary personal growth and transformation.

Sources:

"Public Health Implications of Same-Sex Marriage," *Am J Public Health*

Dana Sourcebook of Brain Science, Third Edition

ZerotoThree.org

www.AThousandMoms.Org

NIH/NIHM "What Is Post Traumatic Stress Disorder, PTSD?"

"PTSD in Children and Adolescents" www.va.gov

http://americanradioworks.publicradio.org/features/romania/b1.html

Cerebrum: The Dana Forum on Brain Science

Chapter Six
Behind the Scenes in the Adolescent Brain

(Book Excerpt). Floyd E. Bloom, M.D., M. Flint Beal, M.D., and David J. Kupfer, M.D., Editors. *Adolescence has been described as a busy time for the human brain. It's a time of transition as the brain, like the rest of the body, physically eases into adulthood and, in the process, the brain's gray matter absorbs an explosion of new external stimuli. In this article, the authors look at the unique external and internal developments of the teenage years: high school, peer pressure, sexuality. The list goes on, and as it does, the brain is challenged. In most cases it thrives; sometimes it does not.*

> ***Several different "transporter" molecules either increase in density during adolescence or reach a plateau.***

A large part of adolescent development takes place in the frontal lobes, which house an incredible number of faculties that we use many times each day. Here are the brain sites that enable us to make sense of the floods of information constantly being gathered by our five senses; to know when we are experiencing an emotion, and even to think about it while we feel it; to understand and keep track of the passage of time; and to hold a thought or object briefly in the forefront of our mind while we proceed with another thought (an ability known as working memory).

According to a recent animal study of frontal lobe development, several different "transporter" molecules, which help the neurons to take in neurotransmitter molecules and break them down for reuse, either increase in density during adolescence or reach a plateau, which in turn alters some signaling pathways and stabilizes others. Partly from refinements in the signal circuits of the frontal lobes and partly through accumulated experience, adolescence gradually brings

SFGate.com
Adolescence requires juggling a busy schedule at school, a host of extra-curricular activities, and a newly complex social schedule.

greater independence along with new capacities to plan, to consider the possible consequences of an action, and to take responsibility for the conduct of one's life. Not surprisingly for a major executive center, the frontal lobes must reorganize to meet new demands, and they do so at more than one level in the years leading up to adulthood. One of the most significant changes (which actually continues well into adulthood) is a major increase in the myelination, or insulation, of the nerve fibers going both into and out of the frontal lobes.

Greater insulation here means faster signaling, and perhaps more highly branched signaling pathways, between frontal lobe neurons and those in any distant region of the brain. This is a development that we can understand on an everyday level. Clearly, the more information the executive center can gather in various modes—visual signals, the emphatic tone of someone's voice, the emotions of the moment—the more nuanced and appropriate the brain's responses can be.

At a day-to-day level, adolescents encounter increasing demands on their attention. For starters, entering middle school or high school means a lot more to keep track of. Instead of being with one teacher in one classroom all day, students move among a half-dozen different classrooms, with a homeroom somewhere else and a locker at yet another place. And, typically today, it quickly becomes necessary to juggle various homework assignments and projects and to balance them against sports or after-school activities, paid or volunteer work, and an ever more complicated social life.

Is it any wonder that researchers, psychologists, and sociologists alike are becoming concerned about the long-term effects of these very crowded schedules on the young, developing brain? Some experts warn that our society may be over encouraging the development of quick responses and mental multitasking in young people, at the expense of equally valuable life skills: planning, thinking things through, and predicting the consequences of actions.

(From The Dana Guide to Brain Health, *Floyd Bloom, M.D., M. Flint Beal, M.D., and David J. Kupfer, M.D., Editors (pp. 104, 105). Copyright © 2003, The Charles A. Dana Foundation. Reprinted with permission of The Free Press, a Division of Simon & Schuster. All rights reserved.)*

Chapter Seven

A Darker Shade of Blue
Teen Depression

Wikimedia Commons
In the digital age, bullying that leads to depression can happen directly or via social media.

Statistics reveal lingering harsh experiences of depression and suicide ideation for youth in middle school and high schools.

Scientists have long acknowledged the brain's circuitry and biochemical processes as integral aspects of depression, specifically as these processes control neurotransmitters that control mood. Beginning in the 1970s, neuroimaging technologies rapidly advanced the study of how brains function – or fail to function. Functional magnetic resonance imaging (fMRI), which became available to researchers over the past 20 years, gives cognitive neuroscientists a 3-D view of neural activity within the brain.

Studies using this technology demonstrate the role of neurotransmitters serotonin, norepinephrine, and dopamine as they regulate mood in the human body. Scientists still aren't exactly sure why individuals with depression have low amounts of these neurotransmitters, yet they do know that for some, antidepressants that specifically target how the brain balances these these neurotransmitters are an effective therapeutic intervention.

Studies show the benefits of combining medication and psychotherapy.

Yet much controversy surrounds the issue of prescribing antidepressants, with some claiming that too often an individual is prescribed a pill without receiving the benefits of psychotherapy or talk therapy. Leigh Matthews, psychologist and director of Urban Psychology in Brisbane, Australia, treats adult clients for depression. There is an abundance of studies evidencing the efficacy of the combination of medication and psychotherapy.

But psychology, Matthews said, tries to first focus on treatment without medication, so it's not always respected by other disciplines, such as general practice physicians or psychiatrists. But the process of psychotherapy and its outcomes last far longer than simply prescribing medication. Yet there are times when medication is absolutely essential, according to Matthews, who also supervises psychologists-in-training at the University of QLD, and those completing their internships through the Australian College of Applied Psychology.

She said when clients are so depressed that they can't get out of bed, think rationally, or use any of the strategies proposed in session, then it's time for medication. Or when clients verbalize suicidal ideation and intent indicating severe depression, then medication is absolutely required.

Also if an individual has a long history of depression, or a strong family history suggesting a genetic basis, then "perhaps they, like a diabetic requiring insulin, need long-term pharmacotherapy to rectify neurochemical imbalances."

Risk Factors for Depression in Gay Teenagers

Gay teenagers not only face the normal physical and emotional stresses of adolescence, but must also contend with developing their sexual identities in a potentially hostile environment.

While straight teenagers also work to develop sexual identities, their peers are generally more accepting of their choices. But gay teenagers often must deal with rejection and teasing because of their sexual orientations.

While society has generally grown more tolerant of homosexuality, gay teenagers still face frequent discrimination and bullying, according to "Victimization of Lesbian, Gay, and Bisexual Youth in a Community Setting."

The article, published in *The Journal of Community Psychology*, examined how frequently gay teenagers experienced rejection, victimization, and other stressors in their lives.

> **Results from the study showed 80 per cent of the gay teens experienced verbal insults while 44 per cent experienced physical assault.**

Researchers Anthony R. D'Augelli and Neil W. Pilkington sampled 194 gay teenagers and surveyed how often they were verbally insulted, physically assaulted, and how their homosexuality affected their family and peer relationships. Results from the study showed 80 per cent of the teens experienced verbal insults because peers knew or thought the teens were gay, while a further 44 per cent experienced physical assault. Additionally, 43 per cent of males and 54 per cent of females said they had lost at least one friend after disclosing their sexuality.

Even though these teens are victimized, they are often fearful to seek help or report bullying because it would reveal their sexual orientations. Their parents might not accept being gay and feel unsympathetic and even might blame the teenager for the bullying. Consider a 14-year-old boy who is a freshman in high school. The boy experiences the same stresses about tests and homework as the other students, but he keeps his biggest source of his stress hidden from his peers, family members, and teachers.

From a young age, the teen knew he was attracted to other boys. Until high school though, he never experienced the kind of bullying he now faces. Now, walking the halls, he's often called a "sissy," "girly-boy," and other more vulgar anti-gay slurs. Soon, the boy wonders if something is wrong with him. He wonders why the other boys in class won't accept him, and if he'll ever be "normal" in their eyes.

He literally has no one to turn to. He hasn't told his parents or close friends his secret, and is afraid that they'll reject him just as his other peers have. The worst part is that the harassment is getting worse, and he isn't sure how to handle the situation anymore. Fearing stigmatization, the boy in the example felt forced to conceal his homosexuality from potential support groups like friends, counselors, and family members. But sometimes, by "coming out" to these groups, some gay teenagers find the help they need to combat negative experiences in high school.

The American Academy of Child and Adolescent Psychiatry lists several concerns of gay teenagers as they develop sexual identities.

Some of these concerns include:

- Feeling alienated from peers
- Feelings of guilt stemming from their orientations
- Worrying about how their parents will respond
- Experiencing teasing and bullying from peers
- Discrimination from clubs or sports
- Experiencing rejection from friends.

Finding Support

Coming out is one of the most stressful periods in a gay teen's life.
In a best-case scenario, the teen's parents might have suspected their teen's sexual orientation, and are happy and accepting of the declaration. But in the worst case, the teen's parents might cut off financial support, reject the teen, and kick him or her out of the household.

In "Homophobic Teasing, Psychological Outcomes, and Sexual Orientation Among High School Students: What Influence Do Parents and Schools Have?" published in *The School Psychology Review,* anti-gay teasing was found to have negative mental health outcomes in gay teenagers who lack supportive school and home environments. Researchers Dorothy L. Esperage and others examined 13,921 high school teenagers, of whom 932 were "questioning" their sexuality, and an additional 1,065 identified themselves as homosexuals.

While all teenagers will experience negative outcomes when parents are unsupportive, for gay teenagers, this effect can be particularly strong, leading to greater instances of depression. The study stated that questioning and openly homosexual students were more likely to report depression, suicide attempts, and drug use when their parents rejected their sexuality.

But gay students who received support were less likely to report these negative outcomes, even if they experienced victimization in school environments. Family acceptance of homosexuality led to higher self-esteem, more support against victimization, and reduced depression among the participants. Since each family scenario differs on a case-by-case basis, the teen should decide if coming out to his

or her parents would cause more harm than good. If so, the teen must identify individuals in the teen's life who will be accepting and supportive of the teen's decision to come out. For some teenagers, this might be an aunt, uncle, or sibling, while others find support in school counselors or close friends.

Sources:

The Journal of Community Psychology

A Thousand Moms

Chapter Eight

Teen Suicide
Death in Life's Springtime

People do not die from suicide. They die due to sadness. --Anonymous

Suicidal feelings are not uncommon. Many people experience transient thoughts of death during a crisis. People under great stress or threat may fleetingly wish they were dead; we call these "passive" suicidal thoughts or passions.

Facebook, courtesy Tammy Aaberg
Justin Aaberg faced unrelenting bullying in his school in Minnesota. He committed suicide at the age of 14.

Experiences with Violence

Negative attitudes toward lesbian, gay, bisexual (LGBT) people put these youth at increased risk for experiences with violence, compared with other students. Violence can include behaviors such as bullying, teasing, harassment, physical assault, and suicide-related behaviors.

Suicidal thoughts or feelings become a cause for more concern if a person experiences them often or starts to consider specific details about ending his or her life. Prolonged or recurring thoughts of suicide are frequently a serious indication of psychiatric illness. And of course, taking any action to commit suicide is a sign that a person needs mental health treatment. Most suicide attempts are not lethal, but are much more than an "emotional cry for help."

Because we often associate suicidal thoughts or behavior with life stresses, we may not realize that the vast majority of suicide victims have an actual psychiatric illness. That condition renders them incapable of coping with stress or loss in healthy ways. Suicidal thoughts are relatively common as a manifestation of clinical depression. A person suffering from severe depression may experience persistent strong impulses to end his or her own life, making specific plans to use highly lethal methods.

Though symptoms of depression are a common thread, there is no one type of person at risk for suicide. Different individuals experience or express their feelings in different ways. Some examples are:

- **Severe hopelessness** coupled with anguish experienced as "psychic pain"; in this condition, a person may suffer in silence, without complaint.
- **A history of substance abuse**, followed by the loss of an important relationship because of this behavior.
- **Severe personality disorder**, which makes people impulsive and very demanding, quick to feel that others do not support them, and unable to soothe their anger over feelings of rejection. If they face a discharge from a hospital, a therapist's vacation, or at the loss of a relationship, they may make repeated suicide attempts, usually by hanging or asphyxiation, escalating the danger each time.

Studies of the neurotransmitter serotonin function that measure receptors in the brains of suicide victims have shown up-regulated--that is hyperactive--receptors, suggesting that serotonin was functioning less than normally in their brains. We do not know if these findings apply to non-suicidal persons with depression.

Other researchers have found increased corticotrophin-releasing hormone (CRH) in the brains of suicide victims. This is important because CRH stimulates the release of adrenocorticotrophin hormone (ACTH) from the pituitary, which in turn stimulates adrenal enlargement and hyperfunction of the adrenal gland, which has been shown to be associated with suicide. Also, CRH may stimulate cells in the brain stem to release norepinephrine, which is associated with emotional arousal. This would fit with doctors' observation of severe anxiety and agitation in many people talking about or attempting suicide.

Recognizing Suicidal Feelings

How do the conditions that can lead to suicide or suicide attempts first appear to others? Depressive illness can begin at any age, but people most commonly show its signs first in young adulthood or adolescence. Often their first symptoms are social withdrawal and a loss of interest in activities that they had previously enjoyed. They may experience persistent insomnia or inability to sleep and sometimes a loss of appetite coupled with weight loss.

> *In a depressed state, people commonly feel increased fatigue, have trouble concentrating, and express hopelessness about life ever improving.*

About half the time clinical depression begins with a major life stress or setback, such as the loss of a relationship or physical illness. In a depressed state, people commonly feel increased fatigue, have trouble concentrating, and express hopelessness about life ever improving. While this condition commonly leads to suicidal feelings, not every depressed person is suicidal, and only a small percentage ever complete a suicide. However, an untreated clinical depression can worsen in severity and lead to such problems as lost jobs, financial stresses and fractured relationships. It's a serious disorder and people of all ages can get help.

How can we tell the difference between depressed patients who are at an immediate risk of suicide and the majority who may not be at risk? This judgment is difficult even for professionals. Clearly expressing a suicidal idea or feelings of hopelessness is an important sign. Other behaviors that indicate a person faces an increased danger of attempting suicide are:

- **Abusing alcohol** or other drugs
- **Impulsive behavior,** such as tantrums, violent outbursts, or episodes of agitation
- **A history of rapid** mood swings between hyperactivity and depression
- **Recurrent severe anxiety,** often in the form of incessant worry and rumination
- **Recurrent panic attacks** in addition to symptoms of depression

Again, not every depressed person who shows one of these signs will commit suicide, but it is cause for extra concern.

It is not uncommon for people who have suicidal feelings to mention them to loved ones or friends. A classic study of suicide in adults reported that over 90 per cent of individuals who committed suicide had a diagnosable psychiatric illness at the time of their death, and more than 60 per cent of those had communicated their suicidal feelings to others--three people, on average--in the year before they died. On the other hand, the same study found that only 18 per cent of this group conveyed their suicidal thoughts to a helping professional, such as a physician or a mental health counselor. This shows the importance of taking a friend's suicidal feelings seriously and reporting them to the person's doctor or therapist. You cannot assume that the doctor or counselor will have heard them.

Teens are not the most likely people to commit suicide (in fact the elderly are), but they have the second highest rate of suicide. Also consider the absolute numbers: while suicide rates are higher in the elderly, the number of deaths by suicide in adolescents is much greater, and 30 per cent higher in LGBT/Q youth. That increased risk, together with the feeling that teenagers have so much potential ahead of them, makes suicide by gay teenagers a great concern for parents, educators, and counselors.

*Organizations such as The Trevor Project serve as lifelines to youth, particularly gay youth. The mission of The Trevor Project is to end suicide among **all** teens, including gay, lesbian, bisexual, transgender and questioning young people. The organization works to fulfill this mission through four strategies:*

- *Provide crisis counseling to young people thinking of suicide.*
- *Offer resources, supportive counseling and a sense of community to LGBTQ young people to reduce the risk that they become suicidal.*
- *Educate young people and adults who interact with young people on LGTBQ-competent suicide prevention, risk detection and response.*
- *Advocate for laws and policies that will reduce suicide among LGBT/Q young people.*

Sources:

A Thousand Moms.org

www.thetrevorproject.org

Dana Sourcebook of Brain Science, Third Edition

Chapter Nine
TBI and Domestic Violence
Healing a Broken Brain

According to an article in *Science Daily*, physicians and researchers at Barrow Neurological Institute have identified a link between domestic violence and traumatic brain injury. The findings could have important implications in the treatment of domestic violence survivors, both in medical and social service communities. The research, led by Dr. Glynnis Zieman, was published in the July 2016 issue of the *Journal of Neurotrauma*.

"Head injuries are among the most common type suffered in domestic violence, which can lead to repetitive brain injuries that often have chronic, life-changing effects, much like what we see in athletes. We found that 88 percent of these victims suffered more than one head injury as a result of their abuse and 81 percent reported too many injuries to count," said Dr. Zieman.

Pixabay

Researchers are uncovering the link between domestic violence and TBI.

The research was conducted at Barrow Concussion and Brain Injury Center in Arizona, where a specialty program has been established to address traumatic brain injury (TBI) in the domestic violence survivor. The program is believed to be the first of its kind in the nation. Dr. Zieman and her team performed a retrospective chart review of more than one hundred patients seen through the program to obtain data for this research.

While concussions have been a significant topic in sports, Barrow has taken special interest in concussions and domestic violence. Barrow experts say that women who previously suffered silently are becoming more aware of the real issue of concussions from their abuse.

The Barrow program provides both medical care and social service assistance for homeless victims who have sustained a TBI as a result of domestic violence. It was created after Barrow social worker Ashley Bridwell and physicians identified a three-way link between homelessness, domestic violence and TBI. The medical team has found many victims are suffering from a full spectrum of side effects that can lead to the loss of a job, income, and eventually homelessness.

"This is the third chapter in the concussion story," says Dr. Zieman. "First it was veterans, then it evolved into professional athletes, and now we're identifying brain injuries in victims of domestic violence. And, unlike well-paid football players, these patients rarely have the support, money and other resources needed to get help."

Annually, TBI injuries cost an estimated $76 billion in direct and indirect medical expenses. The U.S. Centers for Disease Control and Prevention (CDC) statistics for 2010 alone (when the survey was last taken) state:

- TBIs were a factor in the deaths of more than 50,000 people in the United States
- More than 280,000 people with TBI were hospitalized
- 2.2 million people with TBI visited an emergency department

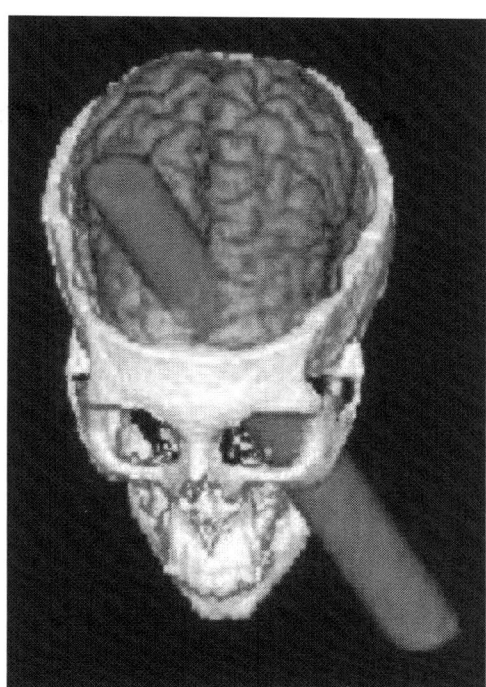

Wikimedia Commons
The case of Phineas Gage uncovered much about the brain and trauma.

This computer-generated graphic shows how, in 1848, a 3-foot long, pointed rod penetrated the skull of Phineas Gage, a railway construction foreman. The rod entered through the side of his face, passed through his brain, and exited his skull. Gage survived the accident but suffered lasting personality and behavioral problems.

Today, we understand a great deal more about the healthy brain and its response to trauma, although science still has much to learn about how to reverse damage resulting from head injuries.

TBI costs the country more than $56 billion a year, and more than 5 million Americans alive today have had a TBI resulting in a permanent need for help in performing daily activities. Survivors of TBI are often left with significant cognitive, behavioral, and communicative disabilities, and some patients develop long-term medical complications, such as epilepsy.

These figures are likely an underestimate of the true number of TBIs as they exclude people who did not seek medical attention at the emergency room. Although approximately 75 percent of brain injuries are considered mild (not life-threatening), as many as 5.3 million people in the United States are estimated to be living with the challenges of long-term TBI-related disability. Not every TBI is alike. Each injury is unique and can cause changes that affect a person for a short period of time, or sometimes permanently.

The majority of people will completely recover from symptoms related to *concussion*, a mild type of TBI. However, persistent symptoms do occur for some people and may last for weeks or months. The long-term effects of TBI may vary depending on the number and nature of "hits" to the head, the age and gender of the individual, the speed with which the person received medical attention, and genetic and other factors.

Over the past few decades preventive measures, such as seatbelts and helmets, and better critical care have substantially increased survival from severe TBI.

> ***TBIs can range from mild (a brief change in mental status or consciousness) to severe (an extended period of unconsciousness or amnesia after the injury).***

Recently, research has expanded from a singular focus on severe TBI to a greater awareness about potential long-term consequences and the need to find better ways to diagnose, treat, and prevent all forms of TBI. Many questions remain unanswered regarding the impact of TBIs, the best treatments, and the most effective methods for promoting recovery of brain function.

What is a Traumatic Brain Injury (TBI)?

A TBI occurs when physical, external forces impact the brain either from a penetrating object or a bump, blow, or jolt to the head. Not all blows or jolts to the head result in a TBI. For the ones that do, TBIs can range from mild (a brief change in mental status or consciousness) to severe (an extended period of unconsciousness or amnesia after the injury). There are two broad types of head injuries: penetrating and non-penetrating.

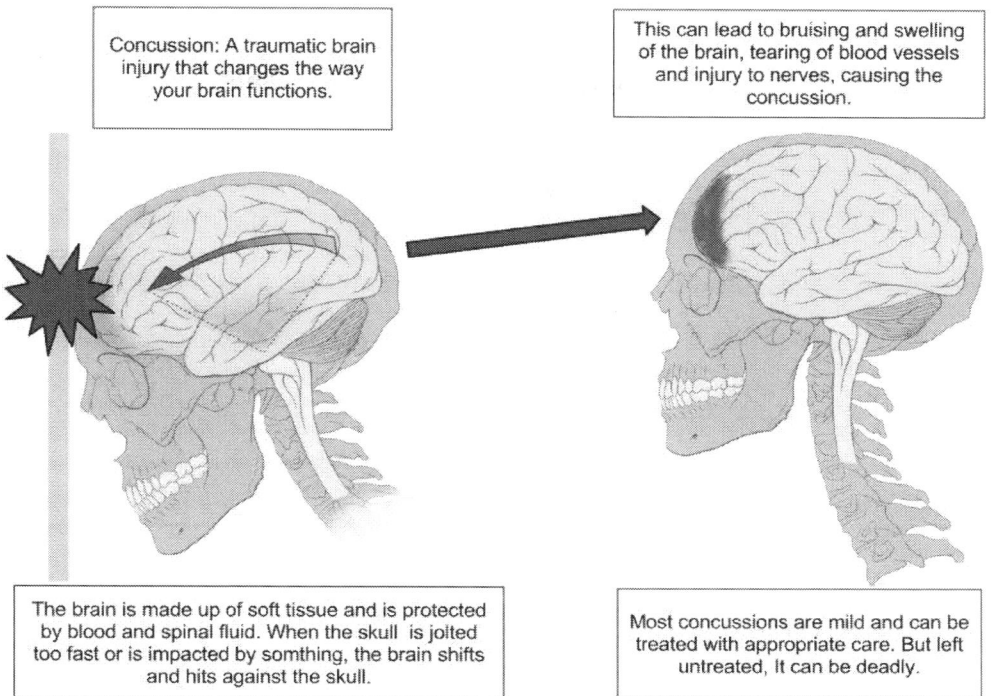

Wikimedia Commons

Penetrating TBI (also known as *open TBI*) occurs when the skull is pierced by an object (for example, a bullet, shrapnel, bone fragment, or by a weapon such as hammer, knife, or baseball bat). With this injury, the object enters the brain tissue.

Non-penetrating TBI (also known as *closed head injury* or *blunt TBI*) is caused by an external force that produces movement of the brain within the skull. Causes include falls, motor vehicle crashes, sports injuries, or being struck by an object. Blast injury due to explosions is a focus of intense study but how it causes brain injury is not fully known.

Some accidents such as explosions, natural disasters, or other extreme events can cause both penetrating and non-penetrating TBI in the same person.

* Terms in italics are defined in the Glossary at the end of this chapter.

How Does TBI Affect the Brain?

TBI-related damage can be confined to one area of the brain, known as a *focal injury*, or it can occur over a more widespread area, known as a *diffuse injury*. The type of injury is another determinant of the effect on the brain. Some injuries are considered *primary*, meaning the damage is immediate.

Other consequences of TBI can be *secondary*, meaning they can occur gradually over the course of hours, days, or weeks. These secondary brain injuries are the result of reactive processes that occur after the initial head trauma.

There are a variety of immediate effects on the brain, including various types of bleeding and tearing forces that injure nerve fibers and cause inflammation, metabolic changes, and brain swelling.

- *Diffuse axonal injury (DAI)* is one of the most common types of brain injuries. DAI refers to widespread damage to the brain's white matter. White matter is composed of bundles of axons (projections of nerve cells that carry electrical impulses). Like the wires in a computer, axons connect various areas of the brain to one another. DAI is the result of *shearing* forces, which stretch or tear these axon bundles. This damage commonly occurs in auto accidents, falls, or sports injuries. It usually results from rotational forces (twisting) or sudden deceleration. It can result in a disruption of neural circuits and a breakdown of overall communication among nerve cells, or *neurons*, in the brain. It also leads to the release of brain chemicals that can cause further damage. These injuries can cause temporary or permanent damage to the brain, and recovery can be prolonged.
- *Concussion* – a type of mild TBI that may be considered a temporary injury to the brain but could take minutes to several months to heal. Concussion can be caused by a number of things including a bump, blow, or jolt to the head, sports injury or fall, motor vehicle accident, weapons blast, or a rapid acceleration or deceleration of the brain within the skull (such as the person having been violently shaken). The individual either suddenly loses consciousness or has sudden altered state of consciousness or awareness, and is often called "dazed" or said to have his/her "bell rung." A second concussion closely following the first one causes further damage to the brain — the so-called "second hit" phenomenon — and can lead to permanent damage or even death in some instances.
- *Hematomas* — a pooling of blood in the tissues outside of the blood vessels. Hematomas can develop when major blood vessels in the head become damaged, causing severe bleeding in and around the brain. Different types of hematomas form depending on where the blood collects relative to the *meninges*. The meninges are the protective membranes surrounding the brain, which consist of three layers: dura mater (outermost), arachnoid mater (middle), and pia mater (innermost).
- *Epidural hematomas* involve bleeding into the area between the skull and the dura mater. These can occur with a delay of minutes to hours after a skull fracture damages an artery under the skull, and are particularly dangerous.

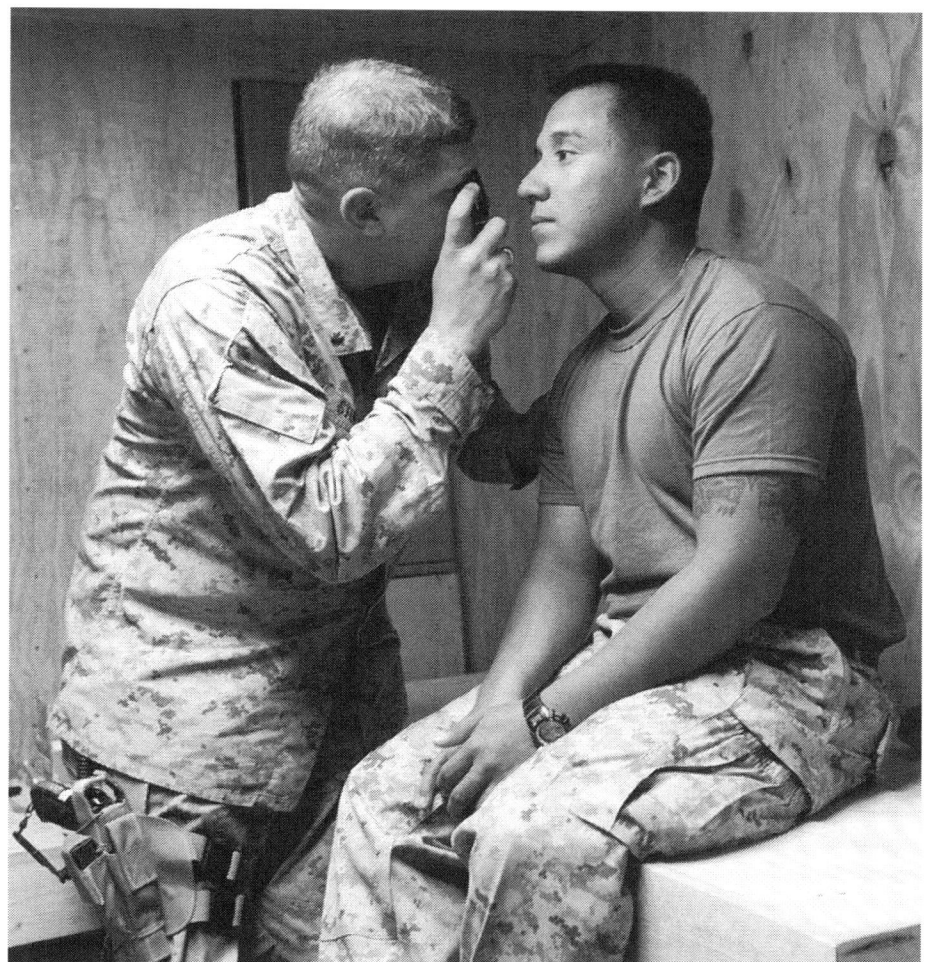
Wikimedia Commons
Soldiers are receiving increased diagnoses and treatment for TBI.

- *Subdural hematomas* involve bleeding between the dura and the arachnoid mater, and like epidural hematomas expert pressure on the outside of the brain . Their effects vary depending on their size and extent to which they compress the brain. They are very common in the elderly after a fall.
- *Subarachnoid hemorrhage* is bleeding that occurs between the arachnoid mater and the pia mater and their effects vary depending on the amount of bleeding.
- Bleeding into the brain itself is called an *intracerebral hematoma* and damages the surrounding tissue.
- *Contusions* — a bruising or swelling of the brain that occurs when very small blood vessels bleed into brain tissue. Contusions can occur directly under the impact site (i.e, a *coup injury*) or, more often, on the complete opposite side of the brain from the impact (i.e., a *contrecoup injury*). They can appear after a delay of hours to a day.

- *Coup/Contrecoup lesions* — contusions or subdural hematomas that occur at the site of head impact as well as directly opposite the coup lesion. Generally they occur when the head abruptly decelerates, which causes the brain to bounce back and forth within the skull (such as in a high-speed car crash). This type of injury also occurs in *shaken baby syndrome*, a severe head injury that results when an infant or toddler is shaken forcibly enough to cause the brain to bounce back and forth against the skull.
- *Skull fractures* — breaks or cracks in one or more of the bones that form the skull. They are a result of blunt force trauma and can cause damage to the underlying areas of the skull such as the membranes, blood vessels, and brain. One main benefit of helmets is to prevent skull fracture.

The first 24 hours after mild TBI are particularly important because subdural hematoma, epidural hematoma, contusion, or excessive brain swelling (edema) are possible and can cause further damage. For this reason doctors suggest watching a person for changes for 24 hours after a concussion.

- *Hemorrhagic progression of a contusion (HPC)* contributes to secondary injuries. HPCs occur when an initial contusion from the primary injury continues to bleed and expand over time. This creates a new or larger lesion — an area of tissue that has been damaged through injury or disease. This increased exposure to blood, which is toxic to brain cells, leads to swelling and further brain cell loss.
- Secondary damage may also be caused by a breakdown in the *blood-brain barrier*. The blood-brain barrier preserves the separation between the brain fluid and the very small capillaries that bring the brain nutrients and oxygen through the blood. Once disrupted, blood, plasma proteins, and other foreign substances leak into the space between neurons in the brain and trigger a chain reaction that causes the brain to swell. It also causes multiple biological systems to go into overdrive, including inflammatory responses which can be harmful to the body if they continue for an extended period of time. It also permits the release of neurotransmitters, chemicals used by brain cells to communicate, which can damage or kill nerve cells when depleted or over-expressed.
- Poor blood flow to the brain can also cause secondary damage. When the brain sustains a powerful blow, swelling occurs just as it would in other parts of the body. Because the skull cannot expand, the brain tissue swells and the pressure inside the skull rises; this is known as *intracranial pressure (ICP)*. When the intracranial pressure becomes too high it prevents blood from flowing to the brain, which deprives it of the oxygen it needs to function. This can permanently damage brain function.

Additional information about TBI and its causes can be found on the U.S. Centers for Disease Control and Prevention TBI website: http://www.cdc.gov/TraumaticBrainInjury/.

What are the leading causes of TBI?

Public Domain Pictures
Car crashes are a leading cause of TBI for adults and children.

Transportation accidents involving automobiles, motorcycles, bicycles, and pedestrians account for half of all TBIs and are the major cause of TBIs in people under age 75.

According to data from the Centers for Disease Control and Prevention (CDC), falls are the most common cause of TBIs and occur most frequently among the youngest and oldest age groups. From 2006 to 2010 alone, falls caused more than half (55 percent) of TBIs among children aged 14 and younger. Among Americans age 65 and older, falls accounted for more than two-thirds (81 percent) of all reported TBIs.

The second and third most common causes of TBI are unintentional blunt trauma (accidents that involved being struck by or against an object), followed closely by motor vehicle accidents. Blunt trauma is especially common in children younger than 15 years old, causing nearly a quarter of all TBIs. Assaults account for an additional 10 percent of TBIs, and include abuse-related TBIs, such as head injuries that result from shaken baby syndrome.

Unintentional blunt trauma includes sports-related injuries, which are also a major cause of TBI. Overall, bicycling, football, playground activities, basketball, and soccer result in the most TBI-related emergency room visits. The cause of these injuries does vary slightly by gender. According to the CDC, among children age 10 to 19, boys are most often injured while playing football or bicycling. Among girls, TBI occur most often while playing soccer or basketball or while bicycling. Anywhere from 1.6 million to 3.8 million sports- and recreation-related TBIs are estimated to occur in the United States annually.

> ***The effects of TBI can range from severe and permanent disability to more subtle functional and cognitive difficulties that often go undetected during initial evaluation.***

TBIs caused by blast trauma from roadside bombs became a common injury to service members in recent military conflicts. From 2000 to 2014 more than 320,000 military service personnel sustained TBIs, though these injuries were not all conflict related. The majority of these TBIs were classified as mild head injuries and due to similar causes as those that occur in civilians.

Adults age 65 and older are at greatest risk for being hospitalized and dying from a TBI, most likely from a fall. TBI-related deaths in children aged 4 years and younger are most likely the result of assault. In young adults aged 15 to 24 years, motor vehicle accidents are the most likely cause. In every age group, serious TBI rates are higher for men than for women. Men are more likely to be hospitalized and are nearly three times more likely to die from a TBI than women.

What are the signs and symptoms of TBI?

The effects of TBI can range from severe and permanent disability to more subtle functional and cognitive difficulties that often go undetected during initial evaluation. These problems may emerge days later. Headache, dizziness, confusion, and fatigue tend to start immediately after an injury, but resolve over time. Emotional symptoms such as frustration and irritability tend to develop later on during the recovery period. Many of the signs and symptoms can be easily missed as people may appear healthy even though they act or feel different. Many of the symptoms overlap with other conditions, such as depression or sleep disorders. If any of the following symptoms appear suddenly or worsen over time following a TBI, especially within the first 24 hours after the injury, people should see a medical professional on an emergency basis

People should seek immediate medical attention if they experience any of the following symptoms:
- loss of or change in consciousness anywhere from a few seconds to a few hours
- decreased level of consciousness, i.e., hard to awaken
- convulsions or seizures
- unequal dilation in the pupils of the eyes or double vision
- clear fluids draining from the nose or ears
- nausea and vomiting
- new neurologic deficit, i.e., slurred speech; weakness of arms, legs, or face; loss of balance

Flickr

Chronic headaches are a major symptom of TBI.

Other common symptoms that should be monitored include:
- mild to profound confusion or disorientation
- problems remembering, concentrating, or making decisions
- headache

- light-headedness, dizziness, vertigo, or loss of balance or coordination
- sensory problems, such as blurred vision, seeing stars, ringing in the ears, bad taste in the mouth
- sensitivity to light or sound
- mood changes or swings, agitation (feeling sad or angry for no reason), combativeness, or other unusual behavior
- feelings of depression or anxiety
- fatigue or drowsiness; a lack of energy or motivation
- changes in sleep patterns (e.g., sleeping a lot more or having difficulty falling or staying asleep); inability to wake up from sleep

Wikimedia
Because of the high popularity of the sport, scrutiny of football players for CTE has caused significant controversy.

Diagnosing TBI in children can be challenging because they may be unable to let others know that they feel different. A child with a TBI may display the following signs or symptoms:
- changes in eating or nursing habits
- persistent crying, irritability, or crankiness; inability to be consoled
- changes in ability to pay attention; lack of interest in a favorite toy or activity
- changes in the way the child plays
- changes in sleep patterns
- sadness or depression
- loss of a skill, such as toilet training
- loss of balance or unsteady walking
- vomiting

In some cases, repeated blows to the head can cause *chronic traumatic encephalopathy (CTE)* – a progressive neurological disorder associated with a variety of symptoms, including cognition and communication problems, motor disorders, problems with impulse control and depression, confusion, and irritability. CTE occurs in those with extraordinary exposure to multiple blows to the head and as a delayed consequence after many years. Studies of retired boxers have shown that repeated blows to the head can cause a number of issues, including memory problems, tremors, and lack of coordination and dementia. Recent studies have demonstrated cases of CTE in other sports with repetitive mild head impacts (e.g., soccer, wrestling, football, and rugby). A single, severe TBI also may lead to a disorder called *post-traumatic dementia (PTD)*, which may be progressive and share some features with CTE. Studies assessing patterns among large populations of people with TBI indicate that moderate or severe TBI in early or mid-life may be associated with increased risk of dementia later in life.

Effects on Consciousness

A TBI can cause problems with arousal, consciousness, awareness, alertness, and responsiveness. Generally, there are four abnormal states that can result from a severe TBI:

- *Brain death* – The lack of measurable brain function and activity after an extended period of time is called brain death and may be confirmed by studies that show no blood flow to the brain.
- *Coma* – A person in a coma is totally unconscious, unaware, and unable to respond to external stimuli such as pain or light. Coma generally lasts a few days or weeks after which an individual may regain consciousness, die, or move into a vegetative state.
- *Vegetative state* – A result of widespread damage to the brain, people in a vegetative state are unconscious and unaware of their surroundings. However, they can have periods of unresponsive alertness and may groan, move, or show reflex responses. If this state lasts longer than a few weeks it is referred to as a *persistent vegetative state*.
- *Minimally conscious state* – People with severely altered consciousness who still display some evidence of self-awareness or awareness of one's environment (such as following simple commands, yes/no responses).

How Is TBI Diagnosed?

Although the majority of TBIs are mild they can still have serious health implications. Of greatest concern are injuries that can quickly grow worse. All TBIs require immediate assessment by a professional who has experience evaluating head injuries. A neurological exam will assess motor and sensory skills and the functioning of one or more cranial nerves. It will also test hearing and speech, coordination and balance, mental status, and changes in mood or behavior, among other abilities. Screening tools for coaches and athletic trainers can identify the most concerning concussions for medical evaluation.

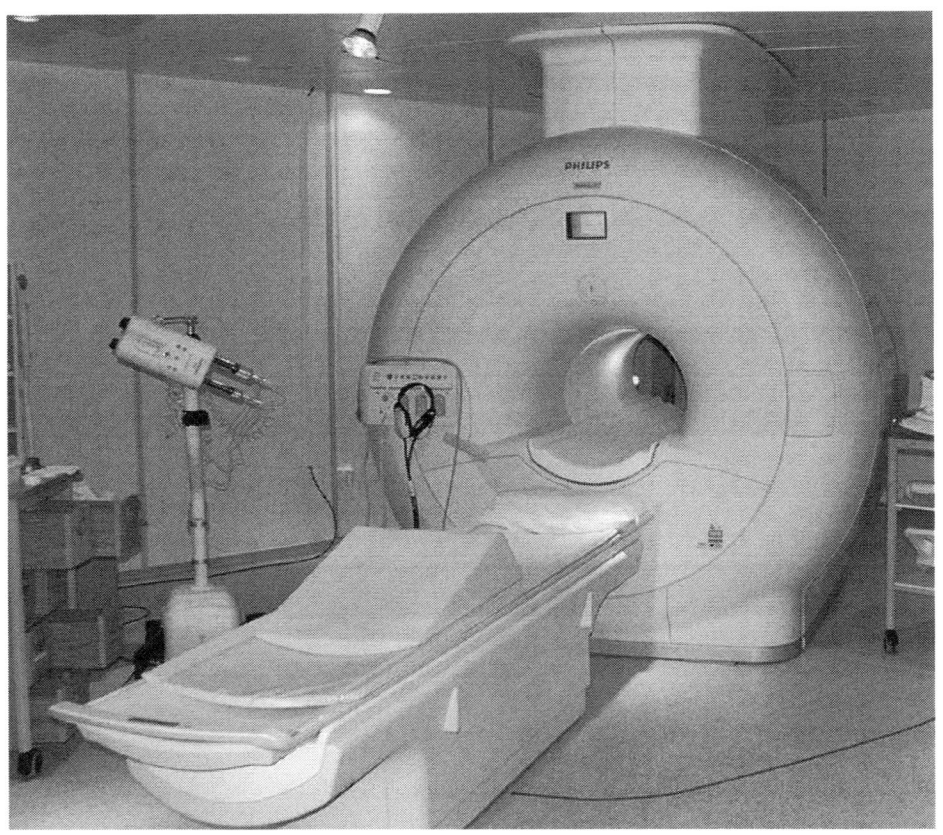

Wikimedia Commons
Brain imaging, including magnetic resonance imaging (MRI), has opened a window on TBI.

Initial assessments may rely on standardized instruments such as the **Acute Concussion Evaluation (ACE)** form from the Centers for Disease Control and Prevention or the **Sport Concussion Assessment Tool 2**, which provide a systematic way to assess a person who has suffered a mild TBI. Reviewers collect information about the characteristics of the injury, the presence of amnesia (loss of memory) and/or *seizures*, as well as the presence of physical, cognitive, emotional, and sleep-related symptoms. The ACE is also used to track symptom recovery over time. It also takes into account risk factors (including concussion, headache, and psychiatric history) that can impact how long it takes to recover from a TBI.

When necessary, medical providers will use brain scans to evaluate the extent of the primary brain injuries and determine if surgery will be needed to help repair any damage to the brain. The need for imaging is based on a physical examination by a doctor and a person's symptoms.

Computed tomography (CT) is the most common imaging technology used to assess people with suspected moderate to severe TBI. CT scans create a series of cross-sectional x-ray images of the skull and brain and can show fractures, hemorrhage, hematomas, hydrocephalus, contusions, and brain tissue swelling. CT scans are often used to assess the damage of a TBI in emergency room settings.

Magnetic resonance imaging (MRI) may be used after the initial assessment and treatment as it is a more sensitive test and picks up subtle changes in the brain that the CT scan might have missed.

The Glasgow Coma Scale is the most widely used tool for assessing the level of consciousness after TBI.

Unlike moderate or severe TBI, milder TBI may not involve obvious signs of damage (hematomas, skull fracture, or contusion) that can be identified with current neuroimaging. Instead, much of what is believed to occur to the brain following mild TBI happens at the cellular level. Significant advances have been made in the last decade to image milder TBI damage. For example, diffusion tensor imaging (DTI) can image white matter tracts, more sensitive tests like fluid-attenuated inversion recovery (FLAIR) can detect small areas of damage, and susceptibility-weighted imaging very sensitively identifies bleeding. Despite these improvements, currently available imaging technologies, blood tests, and other measures remain inadequate for detecting these changes in a way that is helpful for diagnosing the mild concussive injuries.

Neuropsychological tests to gauge brain functioning are often used in conjunction with imaging in people who have suffered mild TBI. Such tests involve performing specific cognitive tasks that help assess memory, concentration, information processing, executive functioning, reaction time, and problem solving. The *Glasgow Coma Scale* is the most widely used tool for assessing the level of consciousness after TBI. The standardized 15-point test measures a person's ability to open his or her eyes and respond to spoken questions or physical prompts for movement. A total score of 3-8 indicates a severe head injury; 9-12 indicates moderate injury; and 13-15 is classified as

mild injury. (For more information about the scale, see
http://glasgowcomascale.org).

Many athletic organizations recommend establishing a baseline picture of an athlete's brain function at the beginning of each season, ideally before any head injuries have occurred. Baseline testing should begin as soon as a child begins a competitive sport. Brain function tests yield information about an individual's memory, attention, and ability to concentrate and solve problems. Brain function tests can be repeated at regular intervals (every 1 to 2 years) and also after a suspected concussion. The results may help health care providers identify any effects from an injury and allow them make more informed decisions about whether a person is ready to return to their normal activities.

How Is TBI Treated?

Many factors, including the size, severity, and location of the brain injury, influence how a TBI is treated and how quickly a person might recover. One of the critical elements to a person's prognosis is the severity of the injury. Although brain injury often occurs at the moment of head impact, much of the damage related to severe TBI develops from secondary injuries which happen days or weeks after the initial trauma. For this reason, people who receive immediate medical attention at a certified trauma center tend to have the best health outcomes.

Treating Mild TBI

Individuals with mild TBI, such as concussion, should focus on symptom relief and "brain rest." In these cases, headaches can often be treated with over-the-counter pain relievers. People with mild TBI are also encouraged to wait to resume normal activities until given permission by a doctor. People with a mild TBI should:
- Make an appointment for a follow-up visit with their health care provider to confirm the progress of their recovery.
- Inquire about new or persistent symptoms and how to treat them.
- Pay attention to any new signs or symptoms even if they seem unrelated to the injury (for example, mood swings, unusual feelings of irritability). These symptoms may be related even if they occurred several weeks after the injury.

Even after symptoms resolve entirely, people should return to their daily activities gradually. Brain functionality may still be limited despite an absence of outward symptoms. Very little is known about the long-term effects of concussions on brain function. There is no clear timeline for a safe return to normal activities although there are guidelines such as those from the American Academy of Neurology and the American Medical Society for Sports Medicine to help determine when athletes can return to practice or competition. Further research is needed to better understand the effects of mild TBI on the brain and to determine when it is safe to resume normal activities.

Preventing future concussions is critical. While most people recover fully from a first concussion within a few weeks, the rate of recovery from a second or third concussion is generally slower.

In the days or weeks after a concussion, a minority of individuals may develop *post-concussion syndrome (PCS)*. People can develop this syndrome even if they never lost consciousness. The symptoms include headache, fatigue, cognitive impairment, depression, irritability, dizziness and balance trouble, and apathy. These symptoms usually improve without medical treatment within one to a few weeks but some people can have longer lasting symptoms.

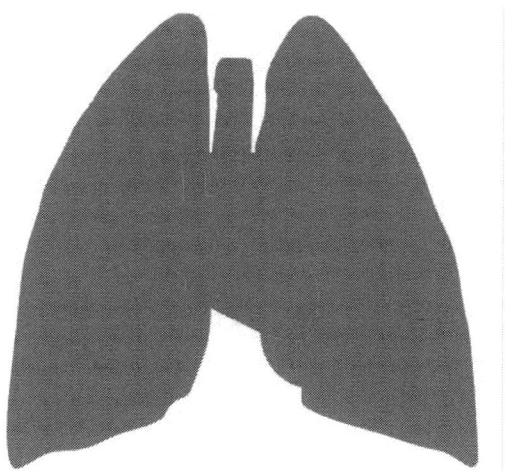

Wikimedia Commons
Vital organs and organ function must be monitored and stabilized in cases of TBI.

In some cases of moderate to severe TBI, persistent symptoms may be related to conditions triggered by imbalances in the production of hormones required for the brain to function normally. Hormone imbalances can occur when certain glands in the body, such as the pituitary gland, are damaged over time as result of the brain injury. Symptoms of these hormonal imbalances include weight loss or gain, fatigue, dry skin, impotence, menstrual cycle changes, depression, difficulty concentrating, hair loss, or cold intolerance. When these symptoms persist 3 months after their initial injury or when they occur up to

3 years after the initial TBI, people should speak with a health care provider about their condition.

Treating Severe TBI

Immediate treatment for the person who has suffered a severe TBI focuses on preventing death; stabilizing the person's spinal cord, heart, lung, and other vital organ functions; and preventing further brain damage. Persons with severe TBI generally require a breathing machine to ensure proper oxygen delivery and breathing.

> *Following the acute care period, people with severe TBI are often transferred to a rehabilitation center.*

During the acute management period, health care providers monitor the person's blood pressure, flow of blood to the brain, brain temperature, pressure inside the skull, and the brain's oxygen supply. A common practice called intracranial pressure ICP monitoring involves inserting a special catheter through a hole drilled into the skull. Doctors frequently rely on ICP monitoring as a way to determine if and when medications or surgery are needed in order to prevent secondary brain injury from swelling. People with severe head injury may require surgery to relieve pressure inside the skull, get rid of damaged or dead brain tissue (especially for penetrating TBI), or remove hematomas.

In-hospital strategies for managing people with severe TBI aim to prevent conditions including:
- Infection, particularly pneumonia
- deep vein thrombosis (blood clots that occur deep within a vein; risk increases during long periods of inactivity)

People with TBIs may need nutritional supplements to minimize the effects that vitamin, mineral, and other dietary deficiencies may cause over time. Some individuals may even require tube feeding to maintain the proper balance of nutrients.

Following the acute care period, people with severe TBI are often transferred to a rehabilitation center where a multidisciplinary team of health care providers help with recovery. The rehabilitation team includes neurologists, nurses, psychologists, nutritionists, as well as physical, occupational, vocational, speech, and respiratory therapists.

Cognitive rehabilitation therapy (CRT) is a strategy aimed at helping individuals regain their normal brain function through an individualized training program. Using this strategy, people may also learn compensatory strategies for coping with persistent deficiencies involving memory, problem solving, and the thinking skills to get things done. CRT programs tend to be highly individualized and their success varies. A 2011 Institute of Medicine report concluded that cognitive rehabilitation interventions need to be developed and assessed more thoroughly.

Other Factors that Influence Recovery

Genes

Evidence suggests that genetics play a role in how quickly and completely a person recovers from a TBI. For example, researchers have found that apolipoprotein E ε4 (ApoE4) — a genetic variant associated with higher risks for Alzheimer's disease — is associated with worse health outcomes following a TBI. Much work remains to be done to understand how genetic factors, as well as how specific types of head injuries in particular locations, affect recovery processes. It is hoped that this research will lead to new treatment strategies and improved outcomes for people with TBI.

Age

Studies suggest that age and the number of head injuries a person has suffered over his or her lifetime are two critical factors that impact recovery. For example, TBI-related brain swelling in children can be very different from the same condition in adults, even when the primary injuries are similar. Brain swelling in newborns, young infants, and teenagers often occurs much more quickly than it does in older individuals. Evidence from very limited CTE studies suggest that younger people (ages 20 to 40) tend to have behavioral and mood changes associated with CTE, while those who are older (ages 50+) have more cognitive difficulties.

Compared with younger adults with the same TBI severity, older adults are likely to have less complete recovery. Older people also have more medical issues and are often taking multiple medications that may complicate treatment (e.g., blood-thinning agents when there is a risk of bleeding into the head). Further research is needed to determine if and how treatment strategies may need to be adjusted based on a person's age.

Researchers are continuing to look for additional factors that may help predict a person's course of recovery.

Can TBI Be Prevented?
The best treatment for TBI is prevention. Unlike most neurological disorders, head injuries can be prevented. According to the CDC, doing the following can help prevent TBIs:
- Wear a seatbelt when you drive or ride in a motor vehicle.
- Wear the correct helmet and make sure it fits properly when riding a bicycle, skateboarding, and playing sports like hockey and football.
- Install window guards and stair safety gates at home for young children.
- Never drive under the influence of drugs or alcohol.
- Improve lighting and remove rugs, clutter, and other trip hazards in the hallway.
- Use nonslip mats and install grab bars next to the toilet and in the tub or shower for older adults.
- Install handrails on stairways.
- Improve balance and strength with a regular physical activity program.
- Ensure children's playgrounds are made of shock-absorbing material, such as hardwood mulch or sand.

Glossary of TBI-related Terms

aneurysm - a blood-filled sac formed by disease related stretching of an artery or blood vessel.
anoxia - an absence of oxygen supply to an organ's tissues leading to cell death.
aphasia - difficulty understanding and/or producing spoken and written language. (*See also* non-fluent aphasia.)
apoptosis - cell death that occurs naturally as part of normal development, maintenance, and renewal of tissues within an organism.
arachnoid membrane - one of the three membranes that cover the brain; it is between the pia mater and the dura. Collectively, these three membranes form the meninges.
brain death - an irreversible cessation of measurable brain function.
Broca's aphasia - *see* non-fluent aphasia.
cerebrospinal fluid (CSF) - the fluid that bathes and protects the brain and spinal cord.
closed head injury - an injury that occurs when the head suddenly and violently hits an object but the object does not break through the skull.
coma - a state of profound unconsciousness caused by disease, injury, or poison.

compressive cranial neuropathies - degeneration of nerves in the brain caused by pressure on those nerves.

computed tomography (CT) - a scan that creates a series of cross-sectional X-rays of the head and brain; also called computerized axial tomography or CAT scan.

concussion - injury to the brain caused by a hard blow or violent shaking, causing a sudden and temporary impairment of brain function, such as a short loss of consciousness or disturbance of vision and equilibrium.

contrecoup - a contusion caused by the shaking of the brain back and forth within the confines of the skull.

contusion - distinct area of swollen brain tissue mixed with blood released from broken blood vessels.

CSF fistula - a tear between two of the three membranes - the dura and arachnoid membranes - that encase the brain.

deep vein thrombosis - formation of a blood clot deep within a vein.

dementia pugilistica - brain damage caused by cumulative and repetitive head trauma; common in career boxers.

depressed skull fracture - a fracture occurring when pieces of broken skull press into the tissues of the brain.

diffuse axonal injury - *see* shearing.

dysarthria - inability or difficulty articulating words due to emotional stress, brain injury, paralysis, or spasticity of the muscles needed for speech.

dura - a tough, fibrous membrane lining the brain; the outermost of the three membranes collectively called the meninges.

early seizures - seizures that occur within 1 week after a traumatic brain injury.

epidural hematoma - bleeding into the area between the skull and the dura.

erosive gastritis - inflammation and degeneration of the tissues of the stomach.

fluent aphasia - a condition in which patients display little meaning in their speech even though they speak in complete sentences. Also called Wernicke's or motor aphasia.

Glasgow Coma Scale - a clinical tool used to assess the degree of consciousness and neurological functioning - and therefore severity of brain injury - by testing motor responsiveness, verbal acuity, and eye opening.

global aphasia - a condition in which patients suffer severe communication disabilities as a result of extensive damage to portions of the brain responsible for language.

hematoma - heavy bleeding into or around the brain caused by damage to a major blood vessel in the head.

hemorrhagic stroke - stroke caused by bleeding out of one of the major arteries leading to the brain.

hypermetabolism - a condition in which the body produces too much heat energy.
hypothyroidism - decreased production of thyroid hormone leading to low metabolic rate, weight gain, chronic drowsiness, dry skin and hair, and/or fluid accumulation and retention in connective tissues.
hypoxia - decreased oxygen levels in an organ, such as the brain; less severe than anoxia.
immediate seizures - seizures that occur within 24 hours of a traumatic brain injury.
intracerebral hematoma - bleeding within the brain caused by damage to a major blood vessel.
intracranial pressure - buildup of pressure in the brain as a result of injury.
ischemic stroke - stroke caused by the formation of a clot that blocks blood flow through an artery to the brain.
locked-in syndrome - a condition in which a patient is aware and awake, but cannot move or communicate due to complete paralysis of the body.
magnetic resonance imaging (MRI) - a noninvasive diagnostic technique that uses magnetic fields to detect subtle changes in brain tissue.
meningitis - inflammation of the three membranes that envelop the brain and spinal cord, collectively known as the meninges; the meninges include the dura, pia mater, and arachnoid.
motor aphasia - *see* non-fluent aphasia.
neural stem cells - cells found only in adult neural tissue that can develop into several different cell types in the central nervous system.
neuroexcitation - the electrical activation of cells in the brain; neuroexcitation is part of the normal functioning of the brain or can also be the result of abnormal activity related to an injury.
neuron - a nerve cell that is one of the main functional cells of the brain and nervous system.
neurotransmitters - chemicals that transmit nerve signals from one neuron to another.
non-fluent aphasia - a condition in which patients have trouble recalling words and speaking in complete sentences. Also called Broca's or motor aphasia.
oligodendrocytes - a type of support cell in the brain that produces myelin, the fatty sheath that surrounds and insulates axons.
penetrating head injury - a brain injury in which an object pierces the skull and enters the brain tissue.
penetrating skull fracture - a brain injury in which an object pierces the skull and injures brain tissue.
persistent vegetative state - an ongoing state of severely impaired consciousness, in which the patient is incapable of voluntary motion.

plasticity - ability of the brain to adapt to deficits and injury.

pneumocephalus - a condition in which air or gas is trapped within the intracranial cavity.

post-concussion syndrome (PCS) - a complex, poorly understood problem that may cause headache after head injury; in most cases, patients cannot remember the event that caused the concussion and a variable period of time prior to the injury.

post-traumatic amnesia (PTA) - a state of acute confusion due to a traumatic brain injury, marked by difficulty with perception, thinking, remembering, and concentration; during this acute stage, patients often cannot form new memories.

post-traumatic dementia - a condition marked by mental deterioration and emotional apathy following trauma.

post-traumatic epilepsy - recurrent seizures occurring more than 1 week after a traumatic brain injury.

prosodic dysfunction - problems with speech intonation or inflection.

pruning - process whereby an injury destroys an important neural network in children, and another less useful neural network that would have eventually died takes over the responsibilities of the damaged network.

seizures - abnormal activity of nerve cells in the brain causing strange sensations, emotions, and behavior, or sometimes convulsions, muscle spasms, and loss of consciousness.

sensory aphasia - *see* fluent aphasia.

shaken baby syndrome - a severe form of head injury that occurs when an infant or small child is shaken forcibly enough to cause the brain to bounce against the skull; the degree of brain damage depends on the extent and duration of the shaking. Minor symptoms include irritability, lethargy, tremors, or vomiting; major symptoms include seizures, coma, stupor, or death.

shearing (or diffuse axonal injury) - damage to individual neurons resulting in disruption of neural networks and the breakdown of overall communication among neurons in the brain.

stupor - a state of impaired consciousness in which the patient is unresponsive but can be aroused briefly by a strong stimulus.

subdural hematoma - bleeding confined to the area between the dura and the arachnoid membranes.

subdural hygroma - a buildup of protein rich fluid in the area between the dura and the arachnoid membranes, usually caused by a tear in the arachnoid membrane.

syndrome of inappropriate secretion of antidiuretic hormone (SIADH) - a condition in which excessive secretion of antidiuretic hormone leads to a sodium deficiency in the blood and abnormally concentrated urine;

symptoms include weakness, lethargy, confusion, coma, seizures, or death if left untreated.

thrombosis or thrombus - the formation of a blood clot at the site of an injury.

vasospasm - exaggerated, persistent contraction of the walls of a blood vessel.

vegetative state - a condition in which patients are unconscious and unaware of their surroundings, but continue to have a sleep/wake cycle and can have periods of alertness.

ventriculostomy - a surgical procedure that drains cerebrospinal fluid from the brain by creating an opening in one of the small cavities called ventricles.

Wernicke's aphasia - *see* fluent aphasia.

Source:

Nih.gov
NIH/NINDS National Institutes of Health/National Institute of Neurological Disorders and Stroke

Chapter Ten

Fighting a National Sleep Crisis
22 Tips for Healthy Rest

Many people view sleep as merely a "down time" when their brains shut off and their bodies rest. People may cut back on sleep, thinking it won't be a problem, because other responsibilities seem much more important. But research shows that a number of vital tasks carried out during sleep help people stay healthy and function at their best. While you sleep, your brain is hard at work forming the pathways necessary for learning and creating memories and new insights. Without enough sleep, you can't focus and pay attention or respond quickly. A lack of sleep may even cause mood problems. Also, growing evidence shows that a chronic lack of sleep increases your risk of obesity, diabetes, cardiovascular disease, and infections.

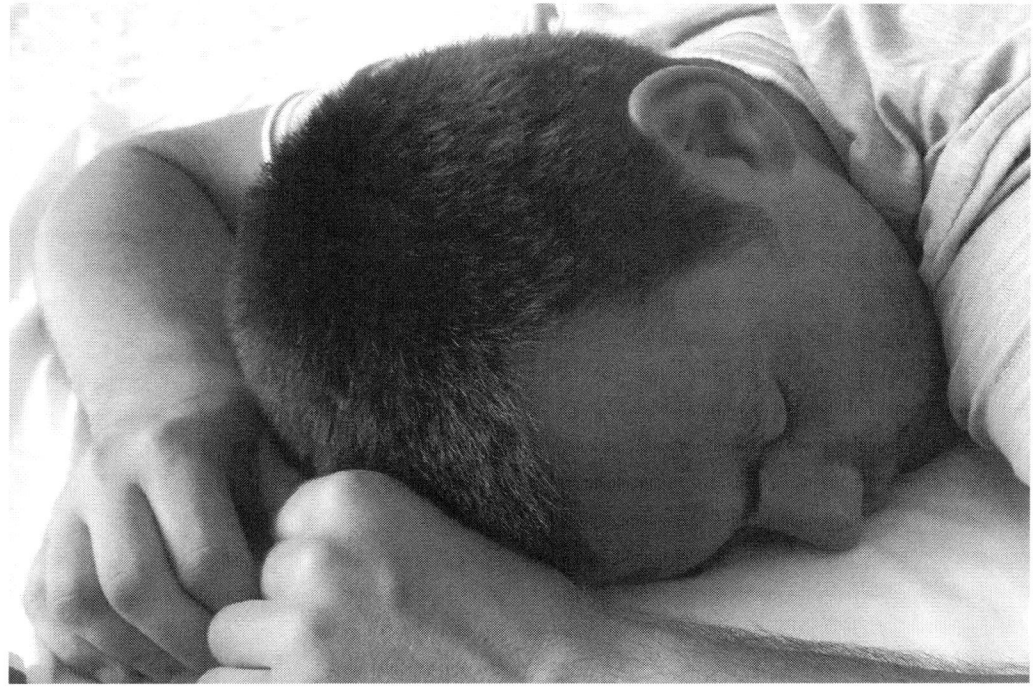

Public Domain Pictures
Researchers acknowledge that regular, consistent sleep plays a major role in brain and body health

Despite growing support for the idea that adequate sleep, like adequate nutrition and physical activity, is vital to our well-being, people are sleeping less. The nonstop "24/7" nature of the world today encourages longer or nighttime work hours and offers continual access to entertainment and other activities. To keep up, people cut back on sleep. A common myth is that people can learn to get by on little sleep (such as less than 6 hours a night) with no adverse effects. Research suggests, however, that adults need at least 7–8 hours of sleep each night to be well rested. Indeed, in 1910, most people slept 9 hours a night. But recent surveys show the average adult now sleeps fewer than 7 hours a night.

Chronic sleep loss or sleep disorders may affect as many as 70 million Americans.

More than one-third of adults report daytime sleepiness so severe that it interferes with work, driving, and social functioning at least a few days each month. Evidence also shows that children's and adolescents' sleep is shorter than recommended. These trends have been linked to increased exposure to electronic media. Lack of sleep may have a direct effect on children's health, behavior, and development. Chronic sleep loss or sleep disorders may affect as many as 70 million Americans. This may result in an annual cost of $16 billion in health care expenses and $50 billion in lost productivity.

What Makes You Sleep? Although you may put off going to sleep in order to squeeze more activities into your day, eventually your need for sleep becomes overwhelming. This need appears to be due, in part, to two substances your body produces. One substance, called adenosine, builds up in your blood while you're awake. Then, while you sleep, your body breaks down the adenosine. Levels of this substance in your body may help trigger sleep when needed.

A buildup of adenosine and many other complex factors might explain why, after several nights of less than optimal amounts of sleep, you build up a sleep debt. This may cause you to sleep longer than normal or at unplanned times during the day. Because of your body's internal processes, you can't adapt to getting less sleep than your body needs. Eventually, a lack of sleep catches up with you. The other substance that helps make you sleep is a hormone called melatonin. This hormone makes you naturally feel sleepy at night. It is part of your internal "biological clock," which controls when you feel sleepy and your sleep patterns. Your biological clock is a small bundle of cells in your brain that works throughout the day and night. Internal and external environmental cues, such as light signals received through your eyes, control these cells. Your biological clock triggers your body to produce melatonin, which helps prepare your brain and body for sleep. As melatonin is released, you'll feel increasingly drowsy.

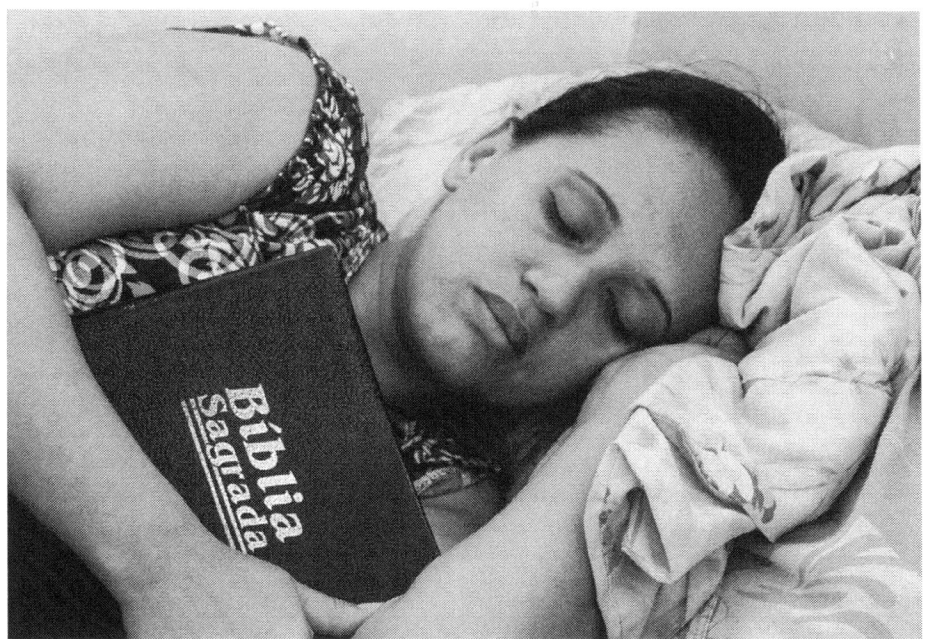

Flickr
Because their work schedules are at odds with powerful sleep-regulating cues like sunlight, night shift workers often find themselves drowsy at work

Because of your biological clock, you naturally feel the most tired between midnight and 7 a.m. You also may feel mildly sleepy in the afternoon between 1 p.m. and 4 p.m. when another increase in melatonin occurs in your body. Your biological clock makes you the most alert during daylight hours and the least alert during the early morning hours. Consequently, most people do their best work during the day.

Our 24/7 society, however, demands that some people work at night. Nearly one-quarter of all workers work shifts that are not during the daytime, and more than two-thirds of these workers have problem sleepiness and/or difficulty sleeping. Because their work schedules are at odds with powerful sleep-regulating cues like sunlight, night shift workers often find themselves drowsy at work, and they have difficulty falling or staying asleep during the daylight hours when their work schedules require them to sleep.

Top 10 Sleep Myths

Myth 1: Sleep is a time when your body and brain shut down for rest and relaxation. No evidence shows that any major organ (including the brain) or regulatory system in the body shuts down during sleep. Some physiological processes actually become more active while you sleep. For example, secretion of certain hormones is boosted, and activity of the pathways in the brain linked to learning and memory increases.

Myth 2: Getting just 1 hour less sleep per night than needed will not have any effect on your daytime functioning. This lack of sleep may not make you noticeably sleepy during the day. But even slightly less sleep can affect your ability to think properly and respond quickly, and it can impair your cardiovascular health and energy balance as well as your body's ability to fight infections, particularly if lack of sleep continues. If you consistently do not get enough sleep, a sleep debt builds up that you can never repay. This sleep debt affects your health and quality of life and makes you feel tired during the day.

Myth 3: Your body adjusts quickly to different sleep schedules. Your biological clock makes you most alert during the daytime and least alert at night. Thus, even if you work the night shift, you will naturally feel sleepy when nighttime comes. Most people can reset their biological clock, but only by appropriately timed cues—and even then, by 1–2 hours per day at best. Consequently, it can take more than a week to adjust to a substantial change in your sleep–wake cycle—for example, when traveling across several time zones or switching from working the day shift to the night shift.

Myth 4: People need less sleep as they get older. Older people don't need less sleep, but they may get less sleep or find their sleep less refreshing. That's because as people age, the quality of their sleep changes. Older people are also more likely to have insomnia or other medical conditions that disrupt their sleep. 23

Myth 5: Extra sleep for one night can cure you of problems with excessive daytime fatigue. Not only is the quantity of sleep important, but also the quality of sleep. Some people sleep 8 or 9 hours a night but don't feel well rested when they wake up because the quality of their sleep is poor. A number of sleep disorders and other medical conditions affect the quality of sleep. Sleeping more won't lessen the daytime sleepiness these disorders or conditions cause. However, many of these disorders or conditions can be treated effectively with changes in behavior or with medical therapies. Additionally, one night of increased sleep may not correct multiple nights of inadequate sleep.

Myth 6: You can make up for lost sleep during the week by sleeping more on the weekends. Although this sleeping pattern will help you feel more rested, it will not completely make up for the lack of sleep or correct your sleep debt. This pattern also will not necessarily make up for impaired performance during the week or the physical problems that can result from not sleeping enough. Furthermore, sleeping later on the weekends can affect your biological clock, making it much harder to go to sleep at the right time on Sunday nights and get up early on Monday mornings.

Myth 7: Naps are a waste of time. Although naps are no substitute for a good night's sleep, they can be restorative and help counter some of the effects of not getting enough sleep at night. Naps can actually help you learn how to do certain tasks quicker. But avoid taking naps later than 3 p.m., particularly if you have trouble falling asleep at night, as late naps can make it harder for you to fall asleep when you go to bed. Also, limit your naps to no longer than 20 minutes, because longer naps will make it harder to wake up and get back in the swing of things. If you take more than one or two planned or unplanned naps during the day, you may have a sleep disorder that should be treated.

Myth 8: Snoring is a normal part of sleep. Snoring during sleep is common, particularly as a person gets older. Evidence is growing that snoring on a regular basis can make you sleepy during the day and increase your risk for diabetes and heart disease. In addition, some studies link frequent snoring to problem behavior and poorer school achievement in children. Loud, frequent snoring also can be a sign of sleep apnea, a serious sleep disorder that should be evaluated and treated. (See "Is Snoring a Problem?" on page 30.)

Myth 9: Children who don't get enough sleep at night will show signs of sleepiness during the day. Unlike adults, children who don't get enough sleep at night typically become hyperactive, irritable, and inattentive during the day. They also have increased risk of injury and more behavior problems, and their growth rate may be impaired. Sleep debt appears to be quite common during childhood and may be misdiagnosed as attention-deficit hyperactivity disorder.

Myth 10: The main cause of insomnia is worry. Although worry or stress can cause a short bout of insomnia, a persistent inability to fall asleep or stay asleep at night can be caused by a number of other factors. Certain medications and sleep disorders can keep you up at night. Other common causes of insomnia are depression, anxiety disorders, and asthma, arthritis, or other medical conditions with symptoms that tend to be troublesome at night. Some people who have chronic insomnia also appear to be more "revved up" than normal, so it is harder for them to fall asleep.

Tips for Getting a Good Night's Sleep

1. **Avoid alcoholic drinks before bed.** Having a "nightcap" or alcoholic beverage before sleep may help you relax, but heavy use robs you of deep sleep and REM sleep, keeping you in the lighter stages of sleep. Heavy alcohol ingestion also may contribute to impairment in breathing at night. You also tend to wake up in the middle of the night when the effects of the alcohol have worn off.

2. **Avoid large meals and beverages late at night.** A light snack is okay, but a large meal can cause indigestion that interferes with sleep.

3. **Drinking too many fluids at night** can cause frequent awakenings to urinate.

4. **If possible, avoid medicines that delay or disrupt your sleep.** Some commonly prescribed heart, blood pressure, or asthma medications, as well as some over-the-counter and herbal remedies for coughs, colds, or allergies, can disrupt sleep patterns. If you have trouble sleeping, talk to your doctor or pharmacist to see whether any drugs you're taking might be contributing to your insomnia and ask whether they can be taken at other times during the day or early in the evening.

5. **Don't take naps after 3 p.m.** Naps can help make up for lost sleep, but late afternoon naps can make it harder to fall asleep at night.

6. **Relax before bed.** Don't overschedule your day so that no time is left for unwinding. A relaxing activity, such as reading or listening to music, should be part of your bedtime ritual. Take a hot bath before bed. The drop in body temperature after getting out of the bath may help you feel sleepy, and the bath can help you relax and slow down so you're more ready to sleep.

7. **Have a good sleeping environment.** Get rid of anything in your bedroom that might distract you from sleep, such as noises, bright lights, an uncomfortable bed, or warm temperatures. You sleep better if the temperature in the room 29 is kept on the cool side. A TV, cell phone, or computer in the bedroom can be a distraction and deprive you of needed sleep.

8. **Having a comfortable mattress and pillow** can help promote a good night's sleep.

9. **Individuals who have insomnia often watch the clock.** Turn the clock's face out of view so you don't worry about the time while trying to fall asleep.

10. **Have the right sunlight exposure.** Daylight is key to regulating daily sleep patterns. Try to get outside in natural sunlight for at least 30 minutes each day. If possible, wake up with the sun or use very bright lights in the morning. Sleep experts recommend that, if you have problems falling asleep, you should get an hour of exposure to morning sunlight and turn down the lights before bedtime.

11. **Don't lie in bed awake.** If you find yourself still awake after staying in bed for more than 20 minutes or if you are starting to feel anxious or worried, get up and do some relaxing activity until you feel sleepy. The anxiety of not being able to sleep can make it harder to fall asleep.

12. **See a doctor if you continue to have trouble sleeping.** If you consistently find it difficult to fall or stay asleep and/ or feel tired or not well rested during the day despite spending enough time in bed at night, you may have a sleep disorder. Your family doctor or a sleep specialist should be able to help you, and it is important to rule out other health or psychiatric problems that may be disturbing your sleep.

Sources:

www.nhlbi.nih.gov

Dana Sourcebook of Brain Science, Third Edition

www.nih.gov

Chapter Eleven

Minority Stress and LGBT/Q Health

The following is adapted from "Public Health Implications of Same-Sex Marriage," Am J Public Health. 2011 June; 101(6): 986–990. William C. Buffie, MD.

One only has to consider the rash of recent teen suicides resulting from anti-gay bullying to begin to comprehend the magnitude of the public health problem faced by this country and its LGBT sexual minority. Despite the prevalence of same-sex households and campaigns to protect human rights, gay persons find the very nature of their being constantly debated within our legislative bodies, the courts, and the mainstream media. They are subject to ridicule and are commonly the targets of demeaning and derogatory slang terms or insensitive jokes. Their morality and value as human beings are frequently questioned by individuals and organizations ignorant or unaccepting of current medical and social science literature concerning the gay population....

Being cast in such a light strongly contributes to the phenomenon known as "minority stress," which members of this community experience in their struggle for validation and acceptance in our heterosexist society.

Wikimedia Commons
To assert and celebrate their community, each year LGBT/Q individuals gather in June at Pride events worldwide.

Unique to the LGBT form of minority stress—as opposed to minority stress engendered by societal prejudice based upon race, ethnicity, gender, or disability—is that one's sexual orientation usually is invisible to others. As a result, in addition to being the target of overt discrimination, LGBT individuals are constantly subject to subtle, inadvertent, or insensitive attacks on the core of their very nature, even by people who profess no disdain or disrespect for them.

For instance, if someone has a lesbian colleague but doesn't know the colleague's orientation, an innocent question—such as asking her if she has a boyfriend, rather than asking "Are you seeing someone special?"—implies a judgment regarding what is "normal." When the "other" is invisible, faceless, or nameless, it is common for those in power to ignore the reality of the other's existence and the challenges the other faces. This interplay of power and prejudice, whether overt or covert, constitutes the phenomenon of heterosexism. Similarities to the racism and sexism so prevalent during the civil rights movements of past generations are obvious.

Internalizing Prejudice

This sexual-minority status, as explained by Riggle and Rostosky, is defined by a culture of devaluation, including overt and subtle prejudice and discrimination, [one that] creates and reinforces the chronic, everyday stress that interferes with optimal human development and well-being.

LGBT individuals, stigmatized by negative societal attitudes directed at the essence of their being, struggle on a daily basis to balance the dual dangers of publicly engaging their need for equality and validation and remaining closeted to find some calm through an escape from public scrutiny. Many gay persons internalize such discrimination and prejudice. Fractured social-support mechanisms and minority-stress–associated low self-esteem contribute to a high prevalence of self-destructive behaviors, such as substance abuse, suicide, and risky sexual behavior.

> ***Institutionalized stigma begets higher rates of sexually transmitted diseases, depression, suicide, and drug use.***

Hatzenbuehler et al. studied more than 34,000 lesbian, gay, and bisexual participants and found empirical evidence of the negative health effects of discriminatory policies relative to marriage equality. They surveyed participants in 2001 and 2002 on a range of psychological health indicators, and they administered the same survey in 2004 and 2005, after 14 states approved constitutional amendments limiting marriage to opposite-sex unions. In the second set of responses, participants reported significantly higher rates of psychiatric disorders, with increases of 36% for any mood disorder, 248% for generalized anxiety disorder, 42% for alcohol use disorder, and 36% for psychiatric comorbidity. In the comparable control group from states without such amendments during the same time period, there were no significant increases in these psychiatric disorders.

Although causality may be difficult to establish, the association and prevalence of these disorders suggest that institutionalized stigma and its attendant internalized prejudice (i.e., minority stress) stand at the forefront of this cycle, begetting higher rates of sexually transmitted diseases, depression, suicide, and drug use—all of which, when combined with suboptimal access to health care and fractured family-support systems, eventually contribute to higher overall mortality as well as morbidity from various cancers, cirrhosis, hypertension, and heart disease....

Chapter Twelve

The Best Little Boy in the World
And Other Coping Strategies Used by Gay Youth

Ever wonder why it's so much easier to remember people's faces than their names? Brain scientists know why. They've identified a pea-size region in the brain that reacts more strongly to faces than it does to cars, dogs, houses or body parts. The evidence is overwhelming that there is a specialized system dedicated to processing faces and not other objects.

The brain's fusiform face area (FFA) sits about halfway back in the head, near the bottom of the visual cortex, the part of the brain than handles vision. The FFA system explains why we are so good at recognizing and remembering faces. A brief glance at a face conveys a wealth of information about identity, expression, gender, age, mood, intent, attractiveness, social states and even honesty.

"The ability to extract this information within a fraction of a second of viewing a face is important for normal social interaction, and has probably played a critical role in the survival of our primate ancestors," writes Dr. Nancy Kanwisher, an investigator at the McGovern Institute for Brain Research at MIT in Cambridge, Mass.

> *A professor and journalist and a gay man, says that ever since he can remember he felt like the character Dr. Richard Kimball in the Fugitive.*

One can imaging gay people possessing among the most highly functioning face-recognition centers.

Dr. Ray Werking, a professor and journalist and a gay man, says that ever since he can remember he felt like the character Dr. Richard Kimball in the TV series and movie, the *Fugitive*. Kimball was wrongly, falsely implicated as a criminal and constantly on the run from nearly everyone, looking forward, backward and always alert to the people around him.

Brain development is literally a lifetime process. In childhood, the brain develops and matures steadily. A child's first six years see the maturing of the auditory cortex (hearing), the visual cortex (seeing), the angular gyrus (understanding languages), and Broca's area (speaking). The prefrontal cortex, site of learning and thinking, matures until age 15 and perhaps beyond.

Adolescence is equally a busy time for the human brain. It's a time of transition as the brain, like the rest of the body, physically eases into adulthood and, in the process, the brain's gray matter absorbs an explosion of new external stimuli. Unique external and internal developments of the teenage years include peer pressure and sexuality. The list goes on, and as it does, the brain is challenged. In most cases it thrives; sometimes it
does not.

Growing Pains in the Teenage Brain

Adolescence marks a turning point of sorts for the brain, as some of its structures are nearing maturity, while others are not yet fully developed. The prefrontal cortex, for example—the brain's center for reason, advance planning, and other higher functions—does not reach maturity until the early 20s. Since this part of the brain seems to act as a kind of cerebral "brake" to halt inappropriate or risky behaviors, some scientists believe sluggish development may explain difficulties in resisting impulsive behavior that some adolescents exhibit at times.

The brain also has ultimate control over the ebb and flow of powerful hormones such as adrenaline, testosterone, and estrogen, which themselves play critical roles in the changing adolescent body. The teenage brain is also struggling to adapt to a shift in the circadian rhythm, the brain's internal biological clock, which drives the sleep-wake cycle. The secretion of melatonin sets the timing for this internal clock, a hormone the brain produces in response to the daily onset of darkness. In one study, researchers found that the further along in puberty teens were, the later at night their melatonin was secreted. In practice, that means teens' natural biological clock is telling them to go to sleep later, and to stay asleep longer.

Wikimedia Commons
A higher visibility for teen lesbian couples enables a sense of living as their true selves.

Matthew Wayne "Matt" Shepard was an American student at the University of Wyoming who was beaten, tortured, and left to die near Laramie, Wyoming on the night of October 6, 1998. He died six days later at Poudre Valley Hospital in Fort Collins, Colorado, on October 12, from severe head injuries.

His death and that of James Byrd in Texas led to the federal enaction of the Shepard/Bird Hate Crimes Act, signed into law by President Barack Obama in 2009.

Matthew Shepard Foundation
The murder of Matthew Shepard in 1998 shocked the world and shone a bright light on bullying of LGBT youth.

In addition to these everyday challenges, at school LGBT students often face harassment, both physical and verbal, which leads to high dropout rates.

- **Gay and transgender** students are two-times less likely to finish high school or pursue a college education compared to the national average.
- **86 percent:** The portion of gay and lesbian students who reported being verbally harassed at school due to their sexual orientation in 2007.
- **44 percent:** The portion of gay and lesbian students who reported being physically harassed at school because of their sexual orientation in 2007.
- **22 percent:** The portion of gay and transgender students who reported having been physically attacked in school in 2007, with 60% saying they did not report the incidents because they believed no one would care.
- **31 percent:** The portion of gay and transgender students who report incidents of harassment and violence at school to staff, only to receive no response.
- **A national study** of students in grades 7-12 found that gay, lesbian, and bisexual youth were more than twice as likely than their heterosexual peers to have attempted suicide and were more likely to have completed suicide.

Despite, or in some cases because of social progress in marriage equality and other matters, pervasiveness of alienation in the lives of the current generation of gay youth is well established. Nevertheless, little is definitively known about the strategies these youth use to cope with stigma and discrimination based on their sexual minority status. Lesbian, gay, bisexual and transgender youth face an array of daunting challenges in addition to many of the developmental stressors facing straight teens.

One of the most difficult stressors gay youth face is heterosexism. This term describes the acculturated and pervasive (intentional or non-intentional) concept that denies, denigrates, and stigmatizes any non-heterosexual form of behavior, identity, relationship, or community. The experience of being stigmatized is at the root of a range of health problems faced by sexual minority adolescents including increased depression, suicide risk, and other mental health disorders. Few researchers have examined the emotional consequences of day-to-day encounters with heterosexism, but many have noted the challenge of maintaining a positive sense of self in the face of chronic negative feedback based in heterosexist attitudes. Recent research has revealed elevated levels of social anxiety in sexual minority adolescents, as well as associations between social anxiety and increased risky sexual behavior.

Forms of heterosexist experiences vary widely, ranging from casual anti-gay remarks to severe physical violence or total social exclusion. Youth encounter heterosexism in diverse settings, including home, school, church, parks, and on the street. Sources of heterosexism were equally wide-ranging, including family members, schoolmates, friends, and religious leaders.

Few researchers have examined the emotional consequences of day-to-day encounters with heterosexism.

Heteroterosexist attitudes by family appeared to be especially stressful for gay youth, in part due to these youths' emotional and financial dependence on their families. The holiday season, the period between Thanksgiving and New Year's Day, accounts for many family disruptions over the issue of a child's sexuality. Youth see this period and the typical gatherings of their nuclear family as an opportunity to begin talking about their true selves. Not coincidentally, this time of year is when the highest incidents of family disruptions occur. Large cities such as San Francisco and New York see an influx of gay youth fleeing their homes and ultimately seeking social services. Relatedly, when television covers an LGBT matter, be it marriage equality or the trauma of the Orlando, FL mass killing at a gay bar, conflicts ensue in families. In 2011, when New York State passed marriage equality, the Ali Forney Center of New York reported a 40 per cent increase in drop-in rates at its youth shelter in New York.

Coping Strategies for Gay Youth
Obtaining information and support through the Internet
For many youth, the Internet served as a means of locating gay-affirmative support that might otherwise have been difficult to obtain. One individual posted poetry about his experiences on a website and received feedback that helped him to increase his sense of self-esteem and reduce feelings of isolation. For this youth, the process of writing poetry had other benefits as well, including cognitively reframing his predicament and venting.

Setting boundaries

A common example of such a strategy involved avoiding individuals who expressed heterosexist attitudes. Youth might stop talking to such a person, or take other active measures to avoid having to encounter them, even if they had formerly been friends.

Some youth express the importance of avoiding heterosexist people, though such strategies could leave them vulnerable to additional psychological, physical and material challenges. For example, leaving home without obtaining alternative sources of support appeared to be a particularly risky means of coping.

> Adam left his small-town home for Los Angeles due to pervasive heterosexism and anti-gay violence he encountered there. He left with

only enough money for train fare and a few essential items, a situation that might have been precarious had he not been able to rely on an aunt living in Los Angeles. He moved in with her and greatly appreciated her support, saying "At least I have one family member that was behind me. But that was the only one." Another youth, who left home for similar reasons had no money whatsoever, but was able to earn income as a dancer in bars. A third respondent, whose brother regularly beat him and called him a "fag," coped by living at friends' homes most of the time.

Passing by telling half-truths

When youth could not avoid the topics that might lead to exposure of their sexual orientation, they often hid their sexual orientation by a careful use of half-truths. For these youth, passing often involved steering a middle course between overt lying and social or familial rejection.

Passing by keeping a low profile in heterosexist environments

Gay people often encounter heterosexist messages in religious settings. Rather than avoid such settings entirely, many respondents continue to attend church, while remaining closeted in that particular environment. In this way, respondents felt they were able to derive benefits from such experiences in spite of hearing heterosexist messages. This was explained by one respondent, who said that he continued to attend church in spite of his discomfort because he valued his relationship with God. Another respondent utilized passing to minimize the embarrassment he and a gay friend would otherwise experience in church when straight men stared at their stereotypically gay attire.

Covering sexual orientation

Some youth who had fully disclosed their sexual orientation nevertheless adopted strategies to minimize its obviousness. Respondents whose families discouraged them from disclosing their sexual orientation to others sometimes used covering as a compromise between their families' wishes and their own. The following respondent recounts an argument in which his parents insisted that he keep his sexual orientation a secret at school.

Listening selectively in stigmatizing environments

Youth often use attentional deployment strategies in religious settings, sometimes by simply ignoring anti-gay messages when they were expressed in church. In order to ignore such heterosexist messages, youth first had to listen to and critically evaluate the ideas they were hearing.

Ignoring provocations

LGBT youth who encounter prejudicial statements directed at them often opted to ignore them. The following respondent reported that when he was younger his mother and sister often said things to him that made him "not feel good about myself". He described that he "used to go so crazy, I yelled at them... like knocked over the TV and stereo, knocked over the whole house." As he got older, he learned to ignore provocations and this helped him to avoid yelling and acting out violently at home. He also stated that he usually uses a similar strategy when encountering prejudice in public settings.

Adopting a self-reliant attitude

Gay youth cope with heterosexism by increasing their personal sense of self-reliance. By cognitively reframing their own circumstances, respondents were able to partially avoid the negative feelings associated with experiences of heterosexist rejection. This process seemed to depend on minimizing the personal relevance of the heterosexist person by discounting the necessity of any support they may previously have provided.

Using media images to support cognitive change

Some youth derived a sense of validation from seeing people on television or in print who shared their sexual minority status. One respondent described feeling more "normal" after seeing gay people represented on television and in magazines. For him, such images included the gay and lesbian television program, *Queer as Folk*, and Matthew Shepard who, though murdered for being gay, is also one of very few nationally-known gay adolescents.

Venting feelings

Expressing emotions by talking or crying was especially important for youth who felt socially isolated, as in the case of one young man, who said, "To deal with sadness, I cried a lot. That relieves the pressure that you have inside." Several respondents also spoke of dealing with strong emotions through creative expression, such as by drawing pictures or writing letters, stories, or poems that conveyed their emotional states. In addition to providing a sense of relief, this strategy enabled respondents to gain insight into their feelings, thus facilitating better cognitive change strategies.

Suppression

Some youth cope with negative emotions by regulating or limiting expression of those emotions. One youth, who experience persistent feelings of sadness, described an encounter in which his best friend confronted him about his suppressive attitude, telling him, "You never express your emotions!" This confrontation resulted in the respondent sharing more of his negative feelings with his friend. At the time of the interview, he continued to use a range of strategies to both diminish negative feelings and avoid expressing them. One therapist commented that many LGBT adults, in his opinion, are beaten up emotionally for being indecisive.

"The Best Little Boy in the World" Syndrome

Recent research reveals why the phrase "the best little boy in the world" aptly describes so many young gay and bisexual men. The phrase derives from the novel published in 1973 by Andrew Tobias, a classic coming out narrative, in which the author recounts his efforts to overcompensate for and evade detection of his nascent sexual orientation by excelling at seemingly everything. Since the publication of Tobias' memoir, numerous gay authors, therapists, and public figures have harnessed the "best little boy in the world" theme to describe their own formative experiences of presenting an infallible facade to guard the personal secret of their sexual orientation.

Wikimedia Commons
Former New Jersey governor Jim McGreevey writes about how the hostilities of his childhood forced him to conceal his sexual orientation.

For example, former New Jersey governor Jim McGreevey in his memoir, *The Confession,* writes about how the hostilities of his childhood environment forced him to conceal his sexual orientation and avidly seek status and achievement instead of same-sex love. He writes, "I think I decided that my ambition would give me more pleasure...than true love."

Downs poignantly discusses the phenomenon of gay men going to great lengths to present an infallible facade to mask their secrets.

Author Paul Monette in his autobiographic account, *Becoming a Man,* describes submerging his gay identity through excelling at school work. Journalist Andrew Sullivan in his cultural commentary, *Love Undetectable,* similarly describes "appeasing my anxiety by perfecting every nook and cranny of my academic requirements." All describe their personal striving to be "the best" in order to cope with their perception of the identity-tarnishing stigma of being gay. The evidence for this phenomenon extends to clinical accounts of gay male development as well.

Psychiatrist Richard Isay writes in *Becoming Gay* that young gay men are forced to become reliant on their own internal resources because approval from others is not guaranteed. Clinical psychologist Alan Downs poignantly discusses the phenomenon of gay men going to great lengths to present an infallible facade to mask their secrets in his clinical account, *The Velvet Rage*. For example, he notes, "We survived by learning to conform to the expectations of others... How would we love ourselves when everything around us told us that we were unlovable? Instead, we chased the affection, approval, and attention doled out by others." Finally, in the *Best Little Boy in the World*, Andrew Tobias notes his early tendency to stave off rejection of his sexual orientation through his academic accomplishments. "Another important line of defense, the most important on a practical day-to-day basis, was my prodigious list of activities...No one could expect me to be out dating...when I had a list of 17 urgent projects to complete."

The key consideration in evaluating the use of chemical substances that can lead to addiction is whether the behavior is to regulate emotional experiences that are already present. Youth in the present study used drugs to both diminish and intensify emotions (to "let everything out," as one youth described his use of LSD). Some respondents also spoke of using substances to cope with feeling isolated.

Using Multiple Strategies to Cope with Heterosexism

Many respondents described a process of experimenting with different coping strategies or adopting multiple approaches based on a range of factors, including the particular form of heterosexism encountered, its setting, its source, or the consequences of using a given strategy.

One especially common combination of strategies involved critically appraising one's current support network.

Sometimes youth adopted a new strategy when the first approach proved ineffective. These youths' coping processes evolved from one set of strategies to another as they discovered new approaches or identified which ones were most effective for their particular circumstances. One especially common combination of strategies involved 1) critically appraising one's current support network, 2) setting boundaries to avoid heterosexist influences, and 3) seeking sources of gay-affirmative social support.

Further research regarding this aspect of how gay and bisexual youth cope with heterosexism is likely to shed additional light on their heroic determination to be themselves in an often unwelcoming world. For such youth, overthinking and silence is unnatural and physically and emotionally taxing at a crucial time.

"The experience of growth and development is challenging enough. For a youth to be burdened with thoughts of day-to-day coping and survival adds unimaginable stress to the process. Youth is a time of creativity, learning, enjoying. Distraction from these necessary parts of growing retards social development, and it will manifest itself soon or during adulthood, when the individual and society pays an increasingly heavy price," says Fred Elia, president of A Thousand Moms: Building Community Support for Lesbian, Gay, Bisexual, and Transgender Youth..

"Well known to professionals in the social welfare community who work with gay youth is the emotional and physical price and lost opportunities suffered by these youth...even those who to one degree or another successfully grow to confident, self-idealized teens and adults. We know that LGBT youth, particularly those who enter foster/adoptive care have three times the likelihood of their straight peers of becoming depressed, succumbing to substance abuse, committing suicide...."

Sources:

J Gay Lesb Soc Serv www.NIH.gov

AThousandMoms.org

Dana Sourcebook of Brain Science

Chapter Thirteen

Substance Abuse
Saying No Is Very Hard to Do

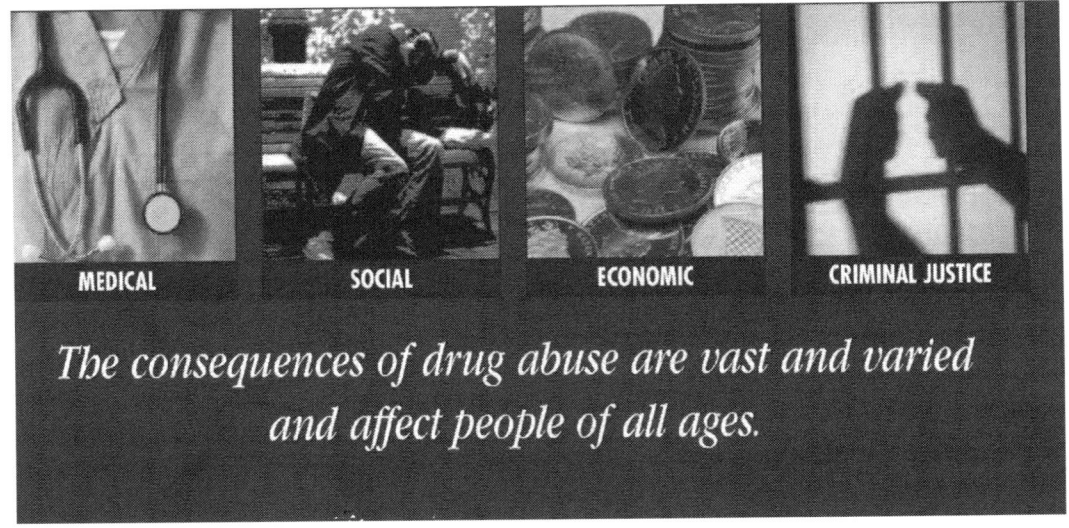

NIDA.gov

Nora D. Volkow, M.D., director of the National Institute on Drug Abuse (www.nida.gov) writes:

"For much of the past century, scientists studying drug abuse labored in the shadows of powerful myths and misconceptions about the nature of addiction. When scientists began to study addictive behavior in the 1930s, people addicted to drugs were thought to be morally flawed and lacking in willpower. Those views shaped society's responses to drug abuse, treating it as a moral failing rather than a health problem, which led to an emphasis on punishment rather than prevention and treatment. Today, thanks to science, our views and our responses to addiction and other substance use disorders have changed dramatically. Groundbreaking discoveries about the brain have revolutionized our understanding of compulsive drug use, enabling us to respond effectively to the problem. As a result of

NIDA.gov
**Nora Volkow, M.D., director,
National Institute on Drug Abuse**

scientific research, we know that addiction is a disease that affects both the brain and behavior. We have identified many of the biological and environmental factors and are beginning to search for the genetic variations that contribute to the development and progression of the disease.

Scientists use this knowledge to develop effective prevention and treatment approaches that reduce the toll drug abuse takes on individuals, families, and communities. Despite these advances, many people today do not understand why people become addicted to drugs or how drugs change the brain to foster compulsive drug use.

Science has revolutionized the understanding of drug addiction.

At the National Institute on Drug Abuse (NIDA), we aim to fill that knowledge gap by providing scientific information about the disease of drug addiction, including the many harmful consequences of drug abuse and the basic approaches that have been developed to prevent and treat substance use disorders. That increased understanding of the basics of addiction will empower people to make informed choices in their own lives, adopt science-based policies and programs that reduce drug abuse and addiction in their communities, and support scientific research that improves the Nation's well-being."

The consequences of drug abuse are vast and varied and affect people of all ages.

I. DRUG ABUSE AND ADDICTION

What is drug addiction? Addiction is defined as a chronic, relapsing brain disease that is characterized by compulsive drug seeking and use, despite harmful consequences. It is considered a brain disease because drugs change the brain—they change its structure and how it works. These brain changes can be long-lasting, and can lead to the harmful behaviors seen in people who abuse drugs.

Addiction is a lot like other diseases, such as heart disease. Both disrupt the normal, healthy functioning of the underlying organ, have serious harmful consequences, and are preventable and treatable, but if left untreated, can last a lifetime.

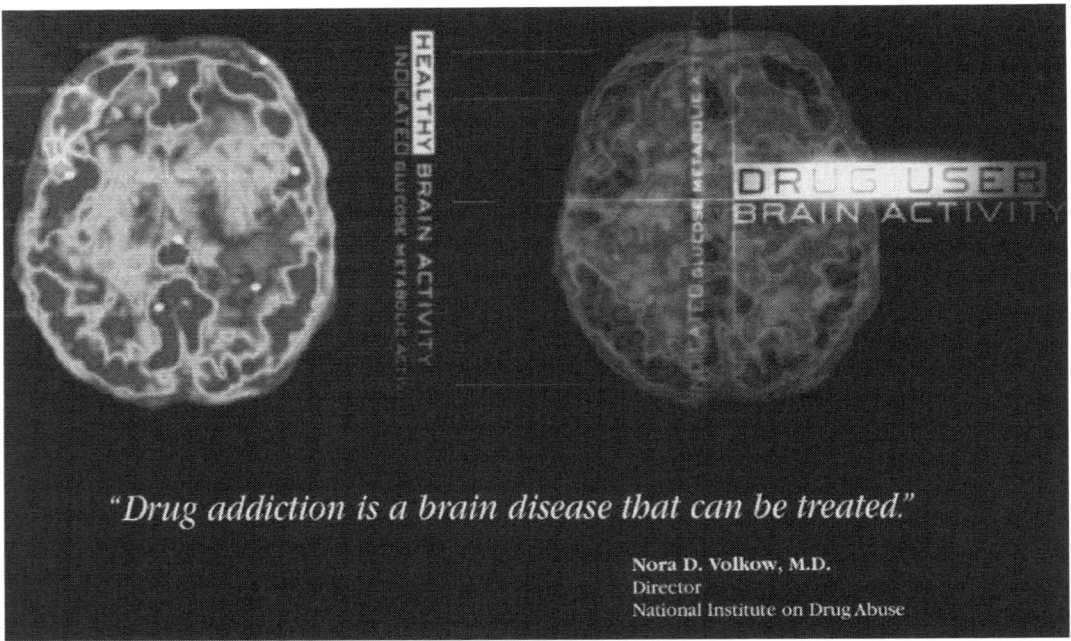

NIDA.gov

Why do people take drugs? In general, people begin taking drugs for a variety of reasons:

- **To feel good.** Most abused drugs produce intense feelings of pleasure. This initial sensation of euphoria is followed by other effects, which differ with the type of drug used. For example, with stimulants such as cocaine, the "high" is followed by feelings of power, self-confidence, and increased energy. In contrast, the

euphoria caused by opiates such as heroin is followed by feelings of relaxation and satisfaction.
- **To feel better.** Some people who suffer from social anxiety, stress-related disorders, and depression begin abusing drugs in an attempt to lessen feelings of distress. Stress can play a major role in beginning drug use, continuing drug abuse, or relapse in patients recovering from addiction.
- **To do better.** Some people feel pressure to chemically enhance or improve their cognitive or athletic performance, which can play a role in initial experimentation and continued abuse of drugs such as prescription stimulants or anabolic/androgenic steroids.
- **Curiosity and "because others are doing it."** In this respect adolescents are particularly vulnerable because of the strong influence of peer pressure. Teens are more likely than adults to engage in risky or daring behaviors to impress their friends and express their independence from parental and social rules.

> *Over time, if drug use continues, other pleasurable activities become less pleasurable, and taking the drug becomes necessary for the user just to feel "normal."*

If taking drugs makes people feel good or better, what's the problem? When they first use a drug, people may perceive what seem to be positive effects; they also may believe that they can control their use. However, drugs can quickly take over a person's life. Over time, if drug use continues, other pleasurable activities become less pleasurable, and taking the drug becomes necessary for the user just to feel "normal." They may then compulsively seek and take drugs even though it causes tremendous problems for themselves and their loved ones.

Some people may start to feel the need to take higher or more frequent doses, even in the early stages of their drug use. These are the telltale signs of an addiction. Even relatively moderate drug use poses dangers. Consider how a social drinker can become intoxicated, get behind the wheel of a car, and quickly turn a pleasurable activity into a tragedy that affects many lives.

Is continued drug abuse a voluntary behavior? The initial decision to take drugs is typically voluntary. However, with continued use, a person's ability to exert self-control can become seriously impaired; this impairment in self-control is the hallmark of addiction. Brain imaging studies of people with addiction show physical changes in areas of the brain that are critical to judgment, decision making, learning and memory, and behavior control. Scientists believe that these changes alter the way the brain works and may help explain the compulsive and destructive behaviors of addiction. Why do some people become addicted to drugs, while others do not?

As with any other disease, vulnerability to addiction differs from person to person, and no single factor determines whether a person will become addicted to drugs. In general, the more risk factors a person has, the greater the chance that individual has of taking drugs.

No single factor determines whether a person will become addicted to drugs.

Protective factors, on the other hand, reduce a person's risk of developing addiction. Risk and protective factors may be either environmental (such as conditions at home, at school, and in the neighborhood) or biological (for instance, a person's genes, their stage of development, and even their gender or ethnicity).

What environmental factors increase the risk of addiction?

Home and Family. The influence of the home environment, especially during childhood, is a very important factor. Parents or older family members who abuse alcohol or drugs, or who engage in criminal behavior, can increase children's risks of developing their own drug problems.

Wikimedia.com
Academic failure or poor social skills can put a child at further risk for using or becoming addicted to drugs.

Peer and School. Friends and acquaintances can have an increasingly strong influence during adolescence. Drug-using peers can sway even those without risk factors to try drugs for the first time. Academic failure or poor social skills can put a child at further risk for using or becoming addicted to drugs.

II. RISK FACTORS

What biological factors increase the risk of addiction? Scientists estimate that genetic factors account for between 40 and 60 percent of a person's vulnerability to addiction; this includes the effects of environmental factors on the function and expression of a person's genes. A person's stage of development and other medical conditions they may have are also factors. Adolescents and people with mental disorders are at greater risk of drug abuse and addiction than the general population.

What other factors increase the risk of addiction?

Early Use. Although taking drugs at any age can lead to addiction, research shows that the earlier a person begins to use drugs, the more likely he or she is to develop serious problems. This may reflect the harmful effect that drugs can have on the developing brain; it also may result from a mix of early social and biological vulnerability factors, including unstable family relationships, exposure to physical or sexual abuse, genetic susceptibility, or mental illness. Still, the fact remains that early use is a strong indicator of problems ahead, including addiction.

Method of Administration.
Smoking a drug or injecting it into a vein increases its addictive potential. Both smoked and injected drugs enter the brain within seconds, producing a powerful rush of pleasure. However, this intense "high" can fade within a few minutes, taking

REDECOM/STEM
Introducing drugs during adolescence may cause brain changes that have profound and long-lasting consequences.

the abuser down to lower, more normal levels. Scientists believe this starkly felt contrast drives some people to repeated drug taking in an attempt to recapture the fleeting pleasurable state.

The brain continues to develop into adulthood and undergoes dramatic changes during adolescence. One of the brain areas still maturing during adolescence is the prefrontal cortex—the part of the brain that enables us to assess situations, make sound decisions, and keep our emotions and desires under control. The fact that this critical part of an adolescent's brain is still a work in progress puts them at increased risk for making poor decisions (such as trying drugs or continuing to take them). Also, introducing drugs during this period of development may cause brain changes that have profound and long-lasting consequences.

III. PREVENTING DRUG ABUSE: THE BEST STRATEGY

Why is adolescence a critical time for preventing drug addiction?

As noted previously, early use of drugs increases a person's chances of developing addiction. Remember, drugs change brains—and this can lead to addiction and other serious problems. So, preventing early use of drugs or alcohol may go a long way in reducing these risks. If we can prevent young people from experimenting with drugs, we can prevent drug addiction.

Wikimedia Commons
In high school, teens may encounter greater availability of drugs, drug use by older teens, and social activities where drugs are used.

Risk of drug abuse increases greatly during times of transition. For an adult, a divorce or loss of a job may lead to drug abuse; for a teenager, risky times include moving or changing schools. In early adolescence, when children advance from elementary through middle school, they face new and challenging social and academic situations. Often during this period, children are exposed to abusable substances such as cigarettes and alcohol for the first time. When they enter high school, teens may encounter greater availability of drugs, drug use by older teens, and social activities where drugs are used.

At the same time, many behaviors that are a normal aspect of their development, such as the desire to try new things or take greater risks, may increase teen tendencies to experiment with drugs. Some teens may give in to the urging of drug-using friends to share the experience with them. Others may think that taking drugs (such as steroids) will improve their appearance or their athletic performance or that abusing substances such as alcohol or MDMA (ecstasy or "Molly") will ease their anxiety in social situations. A growing number of teens are abusing prescription ADHD stimulants such as Adderall® to help them study or lose weight. Teens' still-developing judgment and decision-making skills may limit their ability to accurately assess the risks of all of these forms of drug use. Using abusable substances at this age can disrupt brain function in areas critical to motivation, memory, learning, judgment, and behavior control. So, it is not surprising that teens who use alcohol and other drugs often have family and social problems, poor academic performance, health-related problems (including mental health), and involvement with the juvenile justice system.

IV. DRUGS AND THE BRAIN

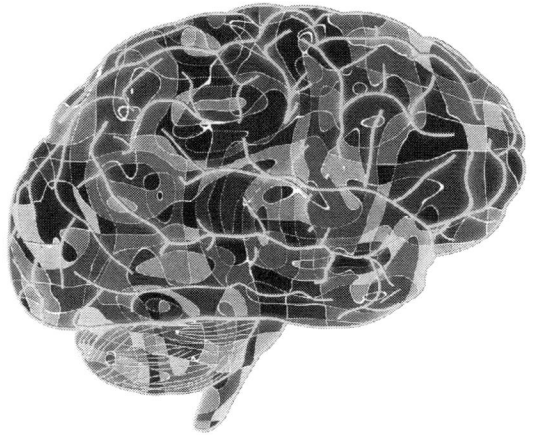

Pixabay.com
The human brain is the most complex organ in the body.

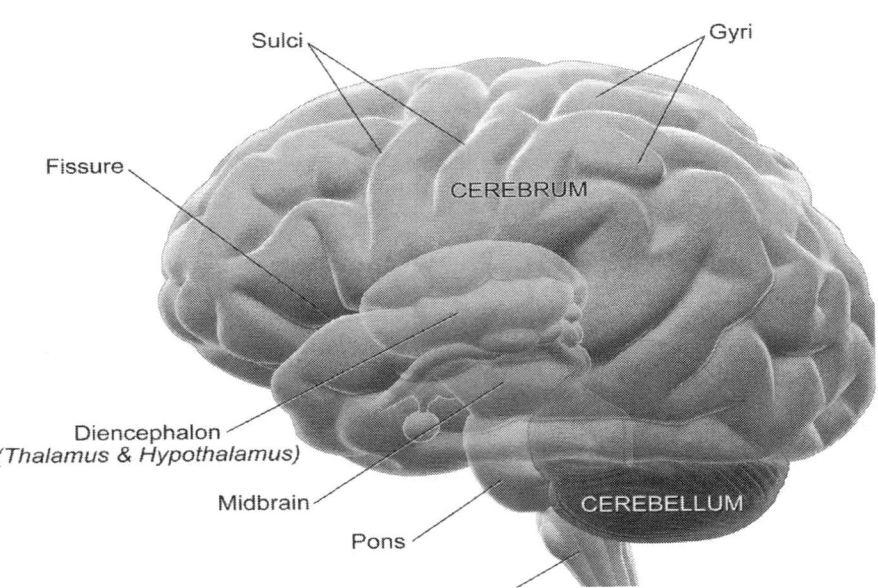

The brain, this three-pound mass of gray and white matter, sits at the center of all human activity—you need it to drive a car, to enjoy a meal, to breathe, to create an artistic masterpiece, and to enjoy everyday activities. In brief, the brain regulates your body's basic functions; enables you to interpret and respond to everything you experience; and shapes your thoughts, emotions, and behavior.

The brain is made up of many parts that all work together as a team. Different parts of the brain are responsible for coordinating and performing specific functions. Drugs can alter important brain areas that are necessary for life-sustaining functions and can drive the compulsive drug abuse that marks addiction.

Key brain areas affected by drug abuse include:
The brain stem, which controls basic functions critical to life, such as heart rate, breathing, and sleeping.
The cerebral cortex, which is divided into areas that control specific functions. Different areas process information from our senses, enabling us to see, feel, hear, and taste. The front part of the cortex, the frontal cortex or forebrain, is the thinking center of the brain; it powers our ability to think, plan, solve problems, and make decisions.
The limbic system, which contains the brain's reward circuit. It links together a number of brain structures that control and regulate our ability to feel pleasure. Feeling pleasure motivates us to repeat behaviors that are critical to our existence. The limbic system is activated by healthy, life-sustaining activities such as eating and socializing— but it is also activated by drugs of abuse. In addition, the limbic system is responsible for our perception of other emotions, both positive and negative, which explains the mood-altering properties of many drugs.

How do drugs work in the brain?

Drugs are chemicals that affect the brain by tapping into its communication system and interfering with the way neurons normally send, receive, and process information. Some drugs, such as marijuana and heroin, can activate neurons because their chemical structure mimics that of a natural neurotransmitter. This similarity in structure "fools" receptors and allows the drugs to attach onto and activate the neurons. Although these drugs mimic the brain's own chemicals, they don't activate neurons in the same way as a natural neurotransmitter, and they lead to abnormal messages being transmitted through the network.

Cigarette smoking among teens is at its lowest point since NIDA began tracking it in 1975. But marijuana use has increased over the past several years as perception of its risks has declined.

Other drugs, such as amphetamine or cocaine, can cause the neurons to release abnormally large amounts of natural neurotransmitters or prevent the normal recycling of these brain chemicals. This disruption produces a greatly amplified message, ultimately disrupting communication channels.

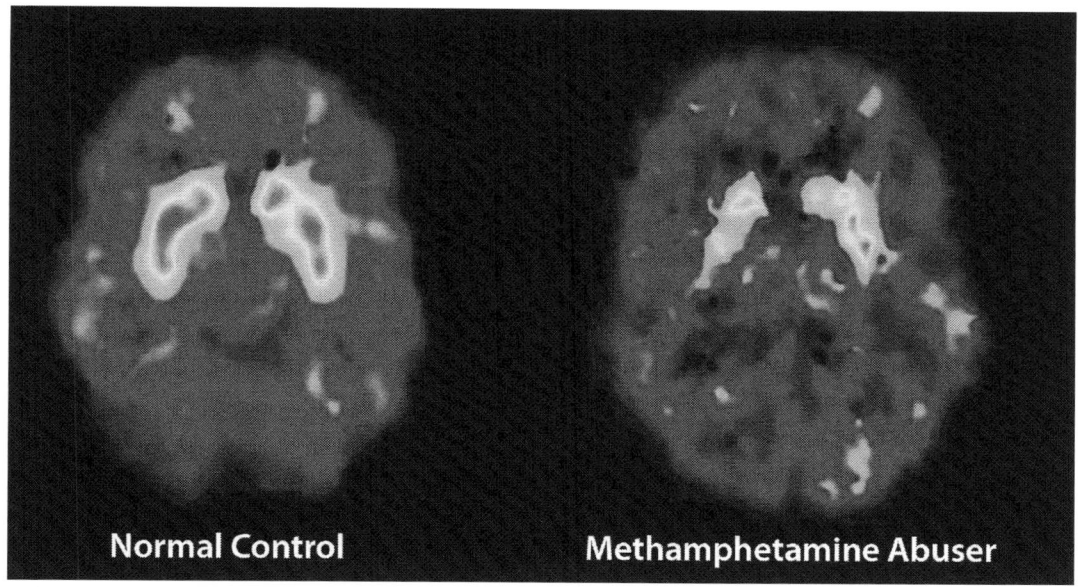

Wikimedia Commons
Brain imaging reveals the diminished brain activity (r) in a substance abuser.

How do drugs work in the brain to produce pleasure? Most drugs of abuse directly or indirectly target the brain's reward system by flooding the circuit with dopamine. Dopamine is a neurotransmitter present in regions of the brain that regulate movement, emotion, motivation, and feelings of pleasure. When activated at normal levels, this system rewards our natural behaviors. Overstimulating the system with drugs, however, produces euphoric effects, which strongly reinforce the behavior of drug use—teaching the user to repeat it.

How does stimulation of the brain's pleasure circuit teach us to keep taking drugs? Our brains are wired to ensure that we will repeat life-sustaining activities by associating those activities with pleasure or reward. Whenever this reward circuit is activated, the brain notes that something important is happening that needs to be remembered, and teaches us to do it again and again without thinking about it. Because drugs of abuse stimulate the same circuit, we learn to abuse drugs in the same way.

Why are drugs more addictive than natural rewards?

When some drugs of abuse are taken, they can release 2 to 10 times the amount of dopamine that natural rewards such as eating and sex do. In some cases, this occurs almost immediately (as when drugs are smoked or injected), and the effects can last much longer than those produced by natural rewards. The resulting effects on the brain's pleasure circuit dwarf those produced by naturally rewarding behaviors. The effect of such a powerful reward strongly motivates people to take drugs again and again. This is why scientists sometimes say that drug abuse is something we learn to do very, very well.

Wikimedia Commons
Abuses of prescription medications and illegal substances create multi-faceted crises in our schools.

What happens to your brain if you keep taking drugs? For the brain, the difference between normal rewards and drug rewards can be described as the difference between someone whispering into your ear and someone shouting into a microphone. Just as we turn down the volume on a radio that is too loud, the brain adjusts to the overwhelming surges in dopamine (and other neurotransmitters) by producing less dopamine or by reducing the number of receptors that can receive signals.

As a result, dopamine's impact on the reward circuit of the brain of someone who abuses drugs can become abnormally low, and that person's ability to experience any pleasure is reduced. This is why a person who abuses drugs eventually feels flat, lifeless, and depressed, and is unable to enjoy things that were previously pleasurable. Now, the person needs to keep taking drugs again and again just to try and bring his or her dopamine function back up to normal—which only makes the problem worse, like a vicious cycle. Also, the person will often need to take larger amounts of the drug to produce the familiar dopamine high—an effect known as tolerance.

How does long-term drug taking affect brain circuits? We know that the same sort of mechanisms involved in the development of tolerance can eventually lead to profound changes in neurons and brain circuits, with the potential to severely compromise the long-term health of the brain. For example, glutamate is another neurotransmitter that influences the reward circuit and the ability to learn. When the optimal concentration of glutamate is altered by drug abuse, the brain attempts to compensate for this change, which can cause impairment in cognitive function. Similarly, long-term drug abuse can trigger adaptations in habit or non-conscious memory systems.

Conditioning is one example of this type of learning, in which cues in a person's daily routine or environment become associated with the drug experience and can trigger uncontrollable cravings whenever the person is exposed to these cues, even if the drug itself is not available. This learned "reflex" is extremely durable and can affect a person who once used drugs even after many years of abstinence.

Drug addiction erodes a person's self-control and ability to make sound decisions, while producing intense impulses to take drugs.

What other brain changes occur with drug abuse? Chronic exposure to drugs of abuse disrupts the way critical brain structures interact to control and inhibit behaviors related to drug use. Just as continued abuse may lead to tolerance or the need for higher drug dosages to produce an effect, it may also lead to addiction, which can drive a user to seek out and take drugs compulsively. Drug addiction erodes a person's self-control and ability to make sound decisions, while producing intense impulses to take drugs.

V. ADDICTION AND HEALTH

What are the medical consequences of drug addiction?
People who suffer from addiction often have one or more accompanying medical issues, which may include lung or cardiovascular disease, stroke, cancer, and mental disorders. Imaging scans, chest X-rays, and blood tests show the damaging effects of long-term drug abuse throughout the body. For example, research has shown that tobacco smoke causes cancer of the mouth, throat, larynx, blood, lungs, stomach, pancreas, kidney, bladder, and cervix.19 In addition, some drugs of abuse, such as inhalants, are toxic to nerve cells and may damage or destroy them either in the brain or the peripheral nervous system.

Does drug abuse cause mental illness or vice versa?
Drug abuse and mental illness often co-exist. In some cases, mental disorders such as anxiety, depression, or schizophrenia may precede addiction; in other cases, drug abuse may trigger or exacerbate those mental disorders, particularly in people with specific vulnerabilities.

How can addiction harm other people?
Beyond the harmful consequences for the person with the addiction, drug abuse can cause serious health problems for others. Three of the more devastating and troubling consequences of addiction are:

1. **Negative effects of prenatal drug exposure on infants and children.** A mother's abuse of heroin or prescription opioids during pregnancy can cause a withdrawal syndrome (called neonatal abstinence syndrome, or NAS) in her infant. It is also likely that some drug exposed children will need educational support in the classroom to help them overcome what may be subtle deficits in developmental areas such as behavior, attention, and thinking. Ongoing research is investigating whether the effects of prenatal drug exposure on the brain and behavior extend into adolescence to cause developmental problems during that time period.

2. **Negative effects of secondhand smoke.** Secondhand tobacco smoke, also called environmental tobacco smoke (ETS), is a significant source of exposure to a large number of substances known to be hazardous to human health, particularly to children. According to the Surgeon General's 2006 Report, The Health Consequences of Involuntary Exposure to Tobacco Smoke, involuntary exposure to secondhand smoke increases the risks of heart disease and lung cancer in people who have never smoked by 25–30 percent and 20–30 percent, respectively.

3. **Increased spread of infectious diseases.** Injection of drugs such as heroin, cocaine, and methamphetamine currently accounts for about 12 percent of new AIDS cases.21 Injection drug use is also a major factor in the spread of hepatitis C, a serious, potentially fatal liver disease. Injection drug use is not the only way that drug abuse contributes to the spread of infectious diseases. All drugs of abuse cause some form of intoxication, which interferes with judgment and increases the likelihood of risky sexual behaviors. This, in turn, contributes to the spread of HIV/AIDS, hepatitis B and C, and other sexually transmitted diseases.

What are some effects of specific abused substances?

Nicotine is an addictive stimulant found in cigarettes and other forms of tobacco. Tobacco smoke increases a user's risk of cancer, emphysema, bronchial disorders, and cardiovascular disease. The mortality rate associated with tobacco addiction is staggering. Tobacco use killed approximately 100 million people during the 20th century, and, if current smoking trends continue, the cumulative death toll for this century has been projected to reach 1 billion.

Alcohol consumption can damage the brain and most body organs. Areas of the brain that are especially vulnerable to alcohol-related damage are the cerebral cortex (largely responsible for our higher brain functions, including problem solving and decision making), the hippocampus (important for memory and learning), and the cerebellum (important for movement coordination).

Marijuana is the most commonly abused illegal substance. This drug impairs short-term memory and learning, the ability to focus attention, and coordination. It also increases heart rate, can harm the lungs, and can increase the risk of psychosis in those with an underlying vulnerability.

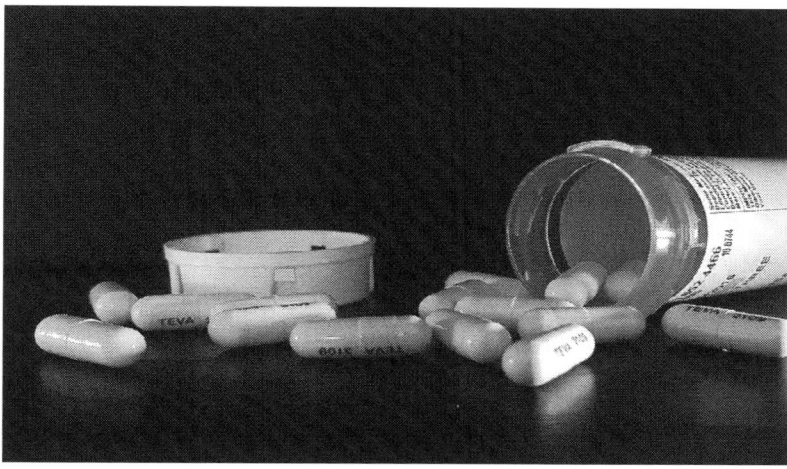

Wikimedia.com
Misuse or abuse of prescription drugs can lead to addiction and even, in some cases, death.

Prescription medications, including opioid pain relievers (such as OxyContin® and Vicodin®), anti-anxiety sedatives (such as Valium® and Xanax®), and ADHD stimulants (such as Adderall® and Ritalin®), are commonly misused to self-treat for medical problems or abused for purposes of getting high or (especially with stimulants) improving performance. However, misuse or abuse of these drugs (that is, taking them other than exactly as instructed by a doctor and for the purposes prescribed) can lead to addiction and even, in some cases, death. Opioid pain relievers, for instance, are frequently abused by being crushed and injected or snorted, greatly raising the risk of addiction and overdose. Unfortunately, there is a common misperception that because medications are prescribed by physicians, they are safe even when used illegally or by another person than they were prescribed for.

Inhalants are volatile substances found in many household products, such as oven cleaners, gasoline, spray paints, and other aerosols, that induce mind-altering effects; they are frequently the first drugs tried by children or young teens. Inhalants are extremely toxic and can damage the heart, kidneys, lungs, and brain. Even a healthy person can suffer heart failure and death within minutes of a single session of prolonged sniffing of an inhalant.

Cocaine is a short-acting stimulant, which can lead users to take the drug many times in a single session (known as a "binge"). Cocaine use can lead to severe medical consequences related to the heart and the respiratory, nervous, and digestive systems.

Amphetamines, including methamphetamine, are powerful stimulants that can produce feelings of euphoria and alertness. Methamphetamine's effects are particularly long-lasting and harmful to the brain. Amphetamines can cause high body temperature and can lead to serious heart problems and seizures.

MDMA (Ecstasy or "Molly") produces both stimulant and mindaltering effects. It can increase body temperature, heart rate, blood pressure, and heart-wall stress. MDMA may also be toxic to nerve cells. z LSD is one of the most potent hallucinogenic, or perception-altering, drugs. Its effects are unpredictable, and abusers may see vivid colors and images, hear sounds, and feel sensations that seem real but do not exist. Users also may have traumatic experiences and emotions that can last for many hours.

Heroin is a powerful opioid drug that produces euphoria and feelings of relaxation. It slows respiration, and its use is linked to an increased risk of serious infectious diseases, especially when taken intravenously. People who become addicted to opioid pain relievers sometimes switch to heroin instead, because it produces similar effects and may be cheaper or easier to obtain.

Steroids, which can also be prescribed for certain medical conditions, are abused to increase muscle mass and to improve athletic performance or physical appearance. Serious consequences of abuse can include severe acne, heart disease, liver problems, stroke, infectious diseases, depression, and suicide.

Drug combinations. A particularly dangerous and common practice is the combining of two or more drugs. The practice ranges from the co-administration of legal drugs, like alcohol and nicotine, to the dangerous mixing of prescription drugs, to the deadly combination of heroin or cocaine with fentanyl (an opioid pain medication). Whatever the context, it is critical to realize that because of drug–drug interactions, such practices often pose significantly higher risks than the already harmful individual drugs.

Nearly half of high school seniors report having used marijuana, and 6.5 percent are daily marijuana users.

VI. TREATMENT AND RECOVERY
Can addiction be treated successfully? YES. Addiction is a treatable disease. Research in the science of addiction and the treatment of substance use disorders has led to the development of evidence-based interventions that help people stop abusing drugs and resume productive lives.

Can addiction be cured? Not always—but like other chronic diseases, addiction can be managed successfully. Treatment enables people to counteract addiction's powerful disruptive effects on their brain and behavior and regain control of their lives.

Newbeginningsteenhelp.com ***Drugs of abuse mimic the natural functioning of the brain.***

Does relapse to drug abuse mean treatment has failed? No. The chronic nature of the disease means that relapsing to drug abuse at some point is not only possible, but likely. Relapse rates (i.e., how often symptoms recur) for people with addiction and other substance use disorders are similar to relapse rates for other well-understood chronic medical illnesses such as diabetes, hypertension, and asthma, which also have both physiological and behavioral components. Treatment of chronic diseases involves changing deeply imbedded behaviors, and relapse does not mean treatment has failed. For a person recovering from addiction, lapsing back to drug use indicates that treatment needs to be reinstated or adjusted or that another treatment should be tried.

What are the principles of effective substance use disorder treatment? Research shows that combining treatment medications (where available) with behavioral therapy is the best way to ensure success for most patients. Treatment approaches must be tailored to address each patient's drug use patterns and drug-related medical, psychiatric, and social problems.

How can medications help treat drug addiction? Different types of medications may be useful at different stages of treatment to help a patient stop abusing drugs, stay in treatment, and avoid relapse.

Treating Withdrawal. When patients first stop using drugs, they can experience a variety of physical and emotional symptoms, including depression, anxiety, and other mood disorders, as well as restlessness or sleeplessness. Certain treatment medications are designed to reduce these symptoms, which makes it easier to stop the drug use.

Staying in Treatment. Some treatment medications are used to help the brain adapt gradually to the absence of the abused drug. These medications act slowly to stave off drug cravings and have a calming effect on body systems. They can help patients focus on counseling and other psychotherapies related to their drug treatment.

Preventing Relapse. Science has taught us that stress, cues linked to the drug experience (such as people, places, things, and moods), and exposure to drugs are the most common triggers for relapse. Medications are being developed to interfere with these triggers to help patients sustain recovery.

How do behavioral therapies treat drug addiction? Behavioral treatments help engage people in substance use disorder treatment, modifying their attitudes and behaviors related to drug use and increasing their life skills to handle stressful circumstances and environmental cues that may trigger intense craving for drugs and prompt another cycle of compulsive use. Behavioral therapies can also enhance the effectiveness of medications and help people remain in treatment longer.

Addiction treatments include:
Tobacco Addiction. Nicotine replacement therapies (available as a patch, inhaler, or gum) • Bupropion • Varenicline
Opioid Addiction • Methadone • Buprenorphine • Naltrexone Alcohol and Drug Addiction • Naltrexone • Disulfiram • Acamprosate
Cognitive Behavioral Therapy seeks to help patients recognize, avoid, and cope with the situations in which they are most likely to abuse drugs.
Contingency Management uses positive reinforcement such as providing rewards or privileges for remaining drug free, for attending and participating in counseling sessions, or for taking treatment medications as prescribed.
Motivational Enhancement Therapy uses strategies to evoke rapid and internally motivated behavior change to stop drug use and facilitate treatment entry.
Family Therapy (especially for youth) approaches a person's drug problems in the context of family interactions and dynamics that may contribute to drug use and other risky behaviors.

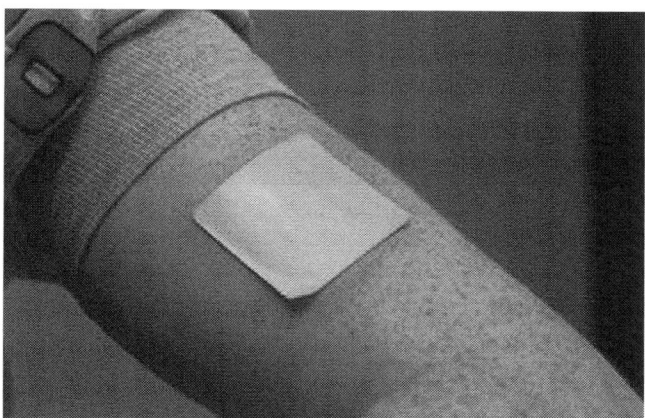

Wikimedia.com ***Use of the nicotine patch has helped many end their cigarette smoking habits.***

Why Do Some Become Addicted While Others Don't?

No single factor can predict whether a person will become addicted to drugs. Risk for addiction is influenced by a combination of factors that include individual biology, social environment, and age or stage of development. The more risk factors an individual has, the greater the chance that taking drugs can lead to addiction. For example:

- **Biology**. The genes that people are born with—in combination with environmental influences—account for about half of their addiction vulnerability. Additionally, gender, ethnicity, and the presence of other mental disorders may influence risk for drug abuse and addiction,
- **Environment**. A person's environment includes many different influences, from family and friends to socioeconomic status and quality of life in general. Factors such as peer pressure, physical and sexual abuse, stress, and quality of parenting can greatly influence the occurrence of drug abuse and the escalation to addiction in a person's life.
- **Development**. Genetic and environmental factors interact with critical developmental stages in a person's life to affect addiction vulnerability. Although taking drugs at any age can lead to addiction, the earlier that drug use begins, the more likely it will progress to more serious abuse, which poses a special challenge to adolescents. Because areas in their brains that govern decision making, judgment, and self-control are still developing, adolescents may be especially prone to risk-taking behaviors, including trying drugs of abuse.

Focus on:
The Heroin Epidemic

Heroin is an illegal, highly addictive drug processed from morphine, a naturally occurring substance extracted from the seed pod of certain varieties of poppy plants. It is typically sold as a white or brownish powder that is "cut" with sugars, starch, powdered milk, or quinine. Pure heroin is a white powder with a bitter taste that predominantly originates in South America and, to a lesser extent, from Southeast Asia, and dominates U.S. markets east of the Mississippi River.

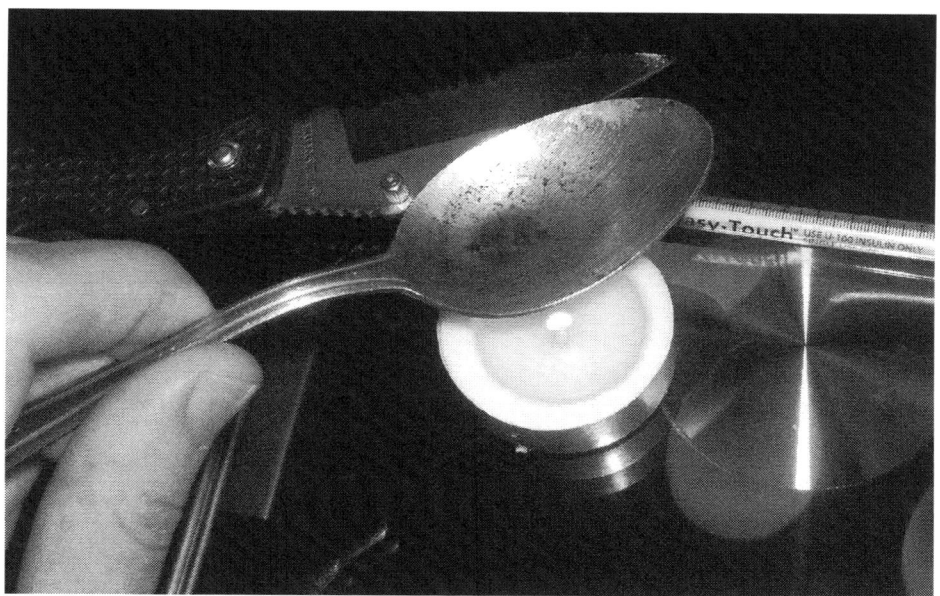

Wikimedia Commons
Increased use of heroin has ravaged families, schools, communities, and a generation of youth.

Highly pure heroin can be snorted or smoked and may be more appealing to new users because it eliminates the stigma associated with injection drug use. "Black tar" heroin is sticky like roofing tar or hard like coal and is predominantly produced in Mexico and sold in U.S. areas west of the Mississippi River.

The dark color associated with black tar heroin results from crude processing methods that leave behind impurities. Impure heroin is usually dissolved, diluted, and injected into veins, muscles, or under the skin.

What are the immediate (short-term) effects of heroin use?

Once heroin enters the brain, it is converted to morphine and binds rapidly to opioid receptors. Abusers typically report feeling a surge of pleasurable sensation—a "rush." The intensity of the rush is a function of how much drug is taken and how rapidly the drug enters the brain and binds to the opioid receptors. With heroin, the rush is usually accompanied by a warm flushing of the skin, dry mouth, and a heavy feeling in the extremities, which may be accompanied by nausea, vomiting, and severe itching. After the initial effects, users usually will be drowsy for several hours; mental function is clouded; heart function slows; and breathing is also severely slowed, sometimes enough to be life-threatening. Slowed breathing can also lead to coma and permanent brain damage.

What are the long-term effects of heroin use?

Repeated heroin use changes the physical structure and physiology of the brain, creating long-term imbalances in neuronal and hormonal systems that are not easily reversed.

Studies have shown some deterioration of the brain's white matter due to heroin use, which may affect decision-making abilities, the ability to regulate behavior, and responses to stressful situations. Heroin also produces profound degrees of tolerance and physical dependence. Tolerance occurs when more and more of the drug is required to achieve the same effects. With physical dependence, the body adapts to the presence of the drug and withdrawal symptoms occur if use is reduced abruptly. Withdrawal may occur within a few hours after the last time the drug is taken. Symptoms of withdrawal include restlessness, muscle and bone pain, insomnia, diarrhea, vomiting, cold flashes with goose bumps ("cold turkey"), and leg movements. Major withdrawal symptoms peak between 24–48 hours after the last dose of heroin and subside after about a week. However, some people have shown persistent withdrawal signs for many months.

Finally, repeated heroin use often results in addiction—a chronic relapsing disease that goes beyond physical dependence and is characterized by uncontrollable drug-seeking no matter the consequences. Heroin is extremely addictive no matter how it is administered, although routes of administration that allow it to reach the brain the fastest (i.e., injection and smoking) increase the risk of addiction. Once a person becomes addicted to heroin, seeking and using the drug becomes their primary purpose in life.

Sources:

National Drug Intelligence Center (2011). The Economic Impact of Illicit Drug Use on American Society. *Washington D.C.: United States Department of Justice.*

Morbidity and Mortality Weekly Report. Available at www.cdc.gov.

Rehm, J., Mathers, C., Popova, S., Thavorncharoensap, M., Teerawattananon Y., Patra, J. *Global burden of disease and injury and economic cost attributable to alcohol use and alcohol-use disorders.* Lancet, 373(9682):2223–2233, 2009.

National Institute on Drug Abuse; National Institutes of Health; U.S. Department of Health and Human Services.

MayoClinic.org

Chapter Fourteen
Fetal Alcohol Spectrum Disorders (FASDs)
A Lifetime of Challenges, Uncertainties

Brain development is most rapid before birth. It maintains a furious pace in infancy and continues briskly through childhood and adolescence, but never ceases altogether. In the third week of gestation, genes switch on to turn some of the embryo's stem cells — "blank slate" cells with the potential to become any kind of tissue — into neurons and glial support cells. These newly formed cells multiply, migrate and connect with one another, guided by chemical signals into the webwork of brain anatomy. By week seven, primitive forms of the cortex, cerebellum and brainstem are apparent.

Birth is only the beginning. The brain adds volume at an initial rate of one per cent per day, growing by two-thirds in the first three months. To fuel its development, the brain requires 43 per cent of the body's daily energy intake until puberty — which, some experts say, explains why physical growth takes so long in humans, compared with other species. Neurons aren't added — in fact, we have more at birth than in adulthood — but grow and connect as specialized circuits form. Sensory centers emerge early, while the hippocampus and amygdala, primitive regions important in emotion and memory, aren't fully functional until age three — which is why we retain virtually no memories of infancy.

> *Of all substances of abuse (including cocaine, heroin, and marijuana), alcohol produces by far the most serious neurobehavioral effects in the fetus.*

Childhood development is a dynamic interaction between the brain and experience. During "critical" periods when regions regulating senses, emotions and language are amped up to make synaptic connections, they must receive appropriate environmental stimulation to connect properly. Development in adolescence defines brain circuits more sharply, adding new synapses, pruning unnecessary ones, and strengthening those that remain.

Sensory, language, and emotional centers mature. Axons add an insulating sheath of myelin to transmit messages more efficiently.

As adolescence ends, the brain still needs fine-tuning, as indicated by frequent risk taking and poor judgment displayed by some in their early 20s. That the prefrontal cortex, the seat of planning and decision making, won't mature fully for another decade partly explains this behavior, but connections between brain regions also must strengthen to give the intellect meaningful control over emotional impulse.

Imaging and research in individuals with Fetal Alcohol Syndrome (FAS) and FASD reveals that some brain regions appear to be most sensitive to prenatal alcohol while other areas apparently are spared adverse effects. Particularly vulnerable regions include the frontal cortex, hippocampus, corpus callosum, and components of the cerebellum.

Pixabay

Any amount of alcohol--wine, beer, or distilled spirits-- passes directly from the mother to the developing baby.

The smallest amount of alcohol--even one glass of wine--passes from the mother to the developing baby. Beer or distilled spirits (vodka, rum, tequila, etc.) all pose risk.

The CDC estimates that 1 in 20 US school children may have FASDs.

Each year, an estimated 7,000 babies are born with prenatal alcohol exposure. Based on the best available data, the CDC Web site states that "experts estimate that the full range of FASDs in the United States and some Western European countries might number as high as 2 to 5 per 100 school children (or 2 per cent to 5 per cent of the population)." The CDC released a fact sheet in 2016 stating that "up to 1 in 20 US school children may have FASDs."

Comprehensive data on the number of individuals with an FASD in the general population of the Unites States, or by state, race or ethnicity, is currently not available.

Fetal Alcohol Spectrum Disorders (FASDs) last a lifetime. There is no cure, but research shows that early intervention treatment services can improve a child's development.

Symptoms of FASDs May Include:

- Abnormal facial features, such as a smooth ridge between the nose and upper lip
- Small head size
- Shorter-than-average height
- Low body weight
- Poor coordination
- Hyperactive behavior
- Difficulty with attention
- Poor memory
- Difficulty in school (especially with math)
- Learning disabilities
- Speech and language delays
- Intellectual disability or low IQ
- Poor reasoning and judgment skills
- Sleep and sucking problems as a baby
- Vision or hearing problems
- Problems with the heart, kidneys, or bones

NOFAS
Treatment for FASDs include medications for some symptoms, behavior and education therapy, parent training, and other alternative approaches.

Types of FASDs

Different terms are used to describe FASDs, depending on the type of symptoms.

Fetal Alcohol Syndrome (FAS): FAS represents the most involved end of the FASD spectrum. Fetal death is the most extreme outcome from drinking alcohol during pregnancy. People with FAS might have abnormal facial features, growth problems, and central nervous system (CNS) problems. People with FAS can have problems with learning, memory, attention span, communication, vision, or hearing. They might have a mix of these problems. People with FAS often have a hard time in school and trouble getting along with others.

Alcohol-Related Neurodevelopmental Disorder (ARND): People with ARND might have intellectual disabilities and problems with behavior and learning. They might do poorly in school and have difficulties with math, memory, attention, judgment, and poor impulse control.

Alcohol-Related Birth Defects (ARBD): People with ARBD might have problems with the heart, kidneys, or bones or with hearing. They might have a mix of these.

There are many types of treatment options, including medication to help with some symptoms, behavior and education therapy, parent training, and other alternative approaches. No one treatment is right for every child. Good treatment plans will include close monitoring, follow-ups, and changes as needed along the way.

FASDs and the Foster Care System

The prevalence of Fetal Alcohol Spectrum Disorders in the foster care system is 10 times higher than in the general population. Eighty per cent of children with FASDs reside in foster or adoptive care. Children in foster care are at higher risk for an FASD. Also, as many as 75 per cent of children in foster care have a family history of mental illness, drug, and/or alcohol abuse. Lack of understanding, frustration and ineffectiveness as a foster parent can lead to multiple placements which increases the childhood trauma for children as they go from one foster placement to the next.

Because of the behaviors and lack of social adaptive skills, children who have been exposed to alcohol prenatally can be challenging to parent. Without understanding of the disability, parents can become overwhelmed and frustrated. Young children with maternal risk factors of substance use, mental health conditions and domestic violence exposure are 2-3 times more likely to experience aggression, anxiety, depression and hyperactivity than children without these maternal risk factors.

A Thousand Moms

Through training and information sharing information on FASDs, foster parents can help reduce unnecessary placements in the child welfare system.

Because so many children who are entering or live in foster care are at risk of FASD, ongoing training on this topic is essential. This training would prevent unnecessary moves within the foster care system and increase identification and diagnosis of this primary disability as well. Informed foster parents lead to informed communities and other professionals.

Helping People Reach Potentials

Also, "protective factors" can help reduce the effects of FASDs and help people with these conditions reach their full potential

Protective factors include: Diagnosis before 6 years of age; a loving, nurturing, and stable home environment during the school year; absence of violence; and involvement in special education and social services.

There are many types of treatment options, including medication to help with some symptoms, behavior and education therapy, parent training, and other alternative approaches. No one treatment is right for every child. Good treatment plans will include close monitoring, follow-ups, and changes as needed along the way.

Because alcohol-related disabilities are widely misdiagnosed and undiagnosed and because FASD is not included in most, if any, health data measures, passive surveillance or clinic-based methods that make no special effort to find FASD but instead rely on treatment or service data from health care facilities or birth records are known to substantially underestimate prevalence rates.

Signs and Symptoms

To diagnose FASDs, doctors look for abnormal facial features (e.g., smooth ridge between nose and upper lip); lower-than-average height, weight, or both; central nervous system problems (e.g., small head size, problems with attention and hyperactivity, poor coordination); and prenatal alcohol exposure (although confirmation is not required to make a diagnosis

The Surgeon General's 2005 Advisory

The office of the United States Surgeon General 2005 Advisory on Alcohol Use in Pregnancy says that a pregnant woman should not drink alcohol during pregnancy. Also, a pregnant woman who already has consumed alcohol during her pregnancy should stop in order to minimize further risk. A woman who is considering becoming pregnant should abstain from alcohol. Health professionals should routinely inquire about alcohol consumption by women of childbearing age, inform them of the risks of alcohol consumption during pregnancy, and advise them not to drink alcoholic beverages during pregnancy. Health professionals may offer brief office-based interventions to women at risk for an alcohol-exposed pregnancy or who are drinking during pregnancy, or may refer them to an alcohol treatment specialist. Finally, according to the advisory, women who continue to have difficulty refraining from alcohol after a brief intervention and those who are alcohol dependent should be referred to an alcohol treatment specialist.

A number of effective tools are available for assessment of at-risk drinking and intervention guidelines for women of childbearing age. Currently, NIH and other agencies and organizations recommend that primary care providers screen all women of childbearing age for alcohol use.

Fetal Alcohol Spectrum Disorders: A Call to Action

Excerpted from Fetal Alcohol Effect. A call to action: Advancing Essential Services and Research on Fetal Alcohol Spectrum Disorders – A report of the National Task Force on Fetal Alcohol Syndrome and Fetal Alcohol Effect, March 2009. Olson H.C., Ohlemiller M.M., O'Connor M.J., Brown C.W., Morris C.A., Damus K., National Task Force on Fetal Alcohol Syndrome. For more information about fetal alcohol spectrum disorders, go to www.cdc.gov/ncbddd/fas

Fetal alcohol spectrum disorders (FASDs) are serious, lifelong birth defects and developmental disabilities caused by prenatal alcohol exposure. They are 100% preventable. Still, a surprisingly large number of children are born with FASDs each year.

FASDs are a public health problem we must face. The U.S. Surgeon General has stated clearly that no amount of alcohol consumption can be considered safe for a pregnant woman and that alcohol can damage a fetus at any stage of pregnancy (Office of the Surgeon General, 2005). Yet, recent U.S. surveys reveal that approximately 12% of pregnant women still drink alcohol (CDC, 2004; SAMHSA, 2007). This means 1 in 8 fetuses are exposed to alcohol and placed at risk for FASDs.

Maternal alcohol use is a growing worldwide phenomenon. It affects children and families of all ethnicities in all societies. Important international collaborative research is beginning to describe the alarming scope of this problem. While community and professional awareness of FASDs have increased, expanded awareness and informed action are sorely needed.

FASDs are considered both medical conditions and developmental disabilities.

FASDs cause a range of lasting medical and developmental problems and result in economic losses of billions of dollars. FASDs can also mean long-standing suffering for families. FASDs are considered both medical conditions and developmental disabilities. They include a wide range of conditions, from subtle neurodevelopmental impairments to the full fetal alcohol syndrome (FAS).

Individuals with FASDs can have physical, mental, behavioral, and/or learning disabilities with possible lifelong implications. Research shows that individuals with FASDs often have significant, long-term deficits in functional life skills. These deficits lead to problems with day-to-day functioning as well as health care, including birth defects and increased risk for injury, unintended pregnancy, and sexually transmitted diseases.

FASDs can also be associated with mental health difficulties, disrupted school and job experiences, trouble with the law, difficulties with independent living, substance abuse, problems with parenting, and more (Bertrand et al., 2004; Streissguth et al., 2004). The median adjusted annual cost of fetal alcohol syndrome has been estimated at $3.6 billion, but the costs associated with the entire fetal alcohol spectrum are surely much higher.

Early, appropriate diagnosis of FASDs is a vital first step to improving outcomes for affected individuals and their families. There is an emerging consensus on how to define FASDs; however, much research is needed to reach a diagnostic standard and to delineate the entire fetal alcohol spectrum.

Diagnostic capacity is growing yet still insufficient in the United States, Canada, and abroad. For this reason, many individuals with FASDs are unrecognized or misdiagnosed. Efforts are ongoing to create and use standardized, reliable diagnostic systems across the globe and to continually improve guidelines as new knowledge emerges from research.

> ***Without suitable treatment and interventions, individuals with FASDs may never reach their full potential.***

Expert opinion from treating professionals, a wealth of family experience, compelling animal research, and pioneering intervention studies indicate that appropriate treatment of FASDs can have a measurable, positive impact. At the present time, even when appropriately diagnosed, individuals with FASDs often receive treatment that is incomplete or inappropriate. Without suitable treatment and interventions, individuals with FASDs may never reach their full potential. Not providing suitable treatment can also result in unnecessary costs as individuals enter systems (such as juvenile justice) with problems that data suggest could have been averted by earlier intervention. Families of affected individuals also need support within the medical and health care systems, and in early intervention and education, juvenile justice and corrections, substance abuse treatment, mental health systems, and social services.

Fortunately, it is possible to define and address the treatment problems raised by FASDs. Because of increasing societal concern, especially over the past 10 years, important steps have been taken and the need for further action made very clear. In the United States, needs assessments have taken place through nationwide public town hall meetings and community agency initiatives (Ryan, Bonnett & Gass, 2006). Intervention guidelines for FASDs are evolving in the United States, Canada, England, and other countries around the world. Strategic research plans are in place to stimulate better description of the entire fetal alcohol spectrum, hone diagnosis, and explain mechanisms of alcohol's action on the developing child's body and brain so that biomarkers and targeted treatments can be identified. An important but limited program of systematic FASD intervention research has begun in the United States and abroad.

Sources:

CDC.gov

Dana Foundation. Dana.org

NOFAS.org

NIH.gov

MOFAS.org

Great Brain Books
In Fiction and Non-Fiction

Originally published as "Madness in Good Company: Great Literary Portrayals of Brain Disorders." From Cerebrum: The Dana Forum on Brain Science, *Vol. 2, No. 3. By Marcia Clendenen and Dick Riley © Dana Press.*

To put these works in a current clinical context, the authors have compared the characteristics and behavior of the characters to the symptoms described in the American Psychiatric Association's *Diagnostic and Statistical Manual of Mental Disorders*, or DSM-IV, the standard descriptive and cataloging text for mental ailments. Some books may be harder to find than others online, in bookstores, or in libraries. If your library does not carry these books, meet with the librarian to request them.

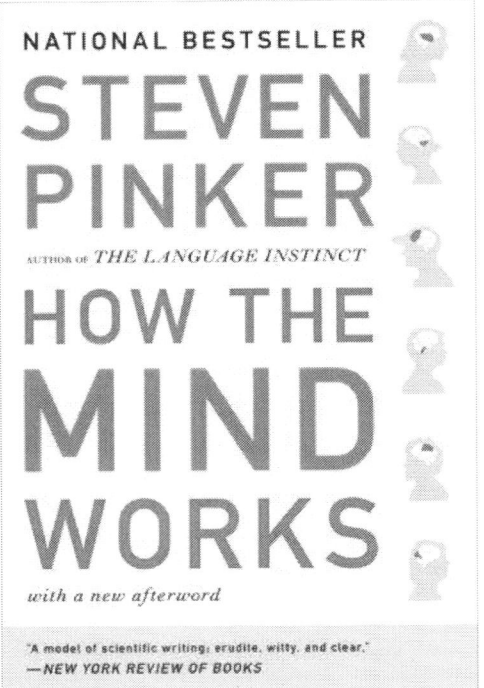

"A Hunger Artist" By Franz Kafka. Fiction. "During these last decades the interest in professional fasting has markedly diminished," begins Franz Kafka in his short story "A Hunger Artist." Those who willingly starve themselves are defined by the DSM-IV as having anorexia nervosa. Certainly the hunger artist exhibits behaviors mentioned in the DSM, including "depressive symptoms such as depressed mood, social withdrawal, irritability, insomnia."

Flowers for Algernon By Daniel Keyes. Fiction. Charlie Gordon is a 32-year-old mentally retarded man who becomes a genius, thanks to a sketchily described new treatment, only to have the process reverse itself. Charlie himself narrates his transformation from a bakery janitor with an intelligence quotient of 68 to a man with an "intelligence that can't really be calculated." The DSM-IV lists both biological and psychosocial factors as potential causes of mental retardation.

Junky By William S. Burroughs. Fiction. "Junk is not a kick. It is a way of life," says Burroughs in his "memoir of a life of addiction." He omits few of the DSM-IV signs of the substance abuser, including "failure to fulfill major role obligations at work, school, or home," and "continued substance use despite having persistent or recurrent...problems caused or exacerbated by the effects of the substance."

Mrs. Dalloway By Virginia Woolf. Fiction. Mrs. Dalloway takes us through one day in the life of Clarissa Dalloway, an upper-class Englishwoman. Her life is contrasted with the tragic story of another major character, Septimus Warren Smith. The novel is set after the end of World War I. Smith had served in the war and suffers from "shell shock," the term then applied to post-traumatic stress disorder.

"Silent Snow, Secret Snow" By Conrad Aiken. Fiction. This short story portrays a 12-year-old boy slipping into an autistic state. Paul avoids the doctor's eyes ("marked impairment in the use of multiple nonverbal behaviors such as eye-to-eye gaze" is one of the DSM's criteria for autism) or else stares, preoccupied with the light in his pupils. Finally he smiles at the secret snow filling the corners of the room.

Slaughterhouse-Five By Kurt Vonnegut. Fiction. This novel tells in a nonlinear narrative of the capture of a young soldier, Billy Pilgrim, and his survival in a meat locker deep below the place where the prisoners are billeted. "Recurrent and intrusive distressing recollections," a classic symptom of post-traumatic stress disorder (PTSD), according to the DSM-IV, could well describe this novel.

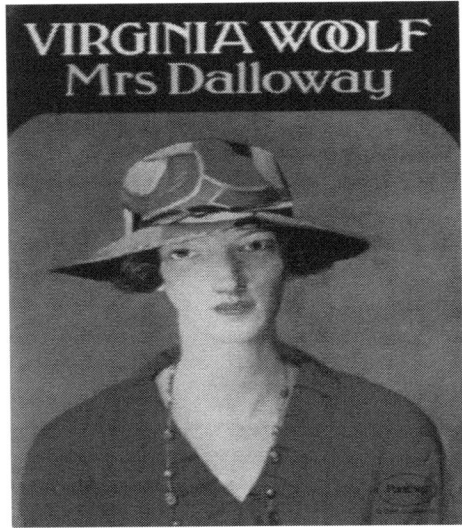 

Tender Is the Night By F. Scott Fitzgerald. Fiction. In *Tender Is the Night*, we observe the tumultuous relationship between Nicole, once diagnosed as schizophrenic, and Dick Diver, who undergoes an alcoholic, downward spiral and professional ruin. Fitzgerald suffered from alcoholism; his wife, Zelda, was diagnosed as schizophrenic and hospitalized. Tender Is the Night was the last novel Fitzgerald completed.

The Accidental Tourist By Anne Tyler. Fiction. Macon Levy's son has been senselessly murdered. Macon's reaction is to create a world of routines, rituals, and dependable habits that hold his grief at bay. The DSM-IV describes obsessive-compulsive disorder as manifesting repetitive behaviors that a person feels driven to perform, behaviors that are aimed—however unrealistically—at preventing or reducing distress or at forestalling some dreaded event or situation.

The Bell Jar Sylvia Plath. Fiction. Esther Greenwood fits many of the DSM-IV criteria for depressive personality disorder. She begins electroshock therapy, but her obsession with thoughts of death worsens. For all its grim subject matter, *The Bell Jar* is full of humor, particularly in its opening passages, and has been described as a female version of the male adolescent rite-of-passage novel, *A Catcher in the Rye*.

The Eden Express By Mark Vonnegut. Books. Non-fiction. This is an autobiographical account of Vonnegut's descent into madness, diagnosed at the time (1970) as schizophrenia. His first psychotic break occurs on a trip and consists of episodes of uncontrolled crying, shaking, and social blunders. This combination of depressive and manic episodes would probably be attributed today to bipolar disorder rather than schizophrenia.

The Idiot By Fyodor Dostoevsky. Fiction. Born in 1821, Dostoevsky became linked with the forces of political reform in Russia. He and a group of friends were arrested for political activity, tried, and sentenced to death. In a dreadful charade, with Dostoevsky already on the scaffold, the sentence was commuted and he was sent to prison in Siberia. There he experienced his first epileptic seizure. Dostoevsky's own epilepsy was particularly acute as he was writing the novel.

The Pickwick Papers By Charles Dickens. Fiction. Among the most memorable of the many comic characters Dickens introduces in *The Pickwick Papers* is Joe, the narcoleptic servant of Mr. Tupman. According to the DSM, narcolepsy— particularly in the extreme form exhibited by Joe— is rare; it would have occurred in as few as 3,600 of the approximately 18 million people in England and Wales at the midpoint of the 19th century.

Adapted from The Great Brain Books Voted by Scientists of the Dana Alliance for Brain Initiatives From Cerebrum: The Dana Forum on Brain Science, *Vol. 1, No. 1, Spring 1999 © Dana Press:*

An Unquiet Mind: A Memoir of Moods and Madness By Kay Redfield Jamison. Vintage. Non-fiction. Jamison, professor of psychiatry at Johns Hopkins University, suffers from manic depression. Here, she reveals how this illness can woo its victims with exalted flights of mind so exhilarating that taking toxic lithium to save their sanity can become an agonizing decision. Jamison makes that issue real for us—in personal, poetic, and scientific terms— as no other writer ever has.

Awakenings By Oliver Sacks. Non-fiction. This Sacks classic is the account of victims of a decades-long sleeping sickness (encephalitis lethargica) who awaken to a new life after being treated with the drug L-dopa. Here, Sacks is able to enter into the world of someone with a neurological disease and help us understand both our common humanity and the medical science.

Descartes' Error: Emotion, Reason, and the Human Brain By Antonio R. Damasio. Non-fiction. The first modern European philosopher, René Descartes, saw mind and body as fundamentally separate. The idea infected Western thought with the premise that rationality and feeling, the mental and the biological, don't mix. Damasio challenges that dualism root and branch, marshaling evidence from basic and clinical research and interpreting it with rare philosophical acuity.

Drugs and the Brain By Solomon H. Snyder. Non-fiction. Snyder tells the story of brain research from the viewpoint of brain chemistry and pharmacological agents (some known over thousands of years) and what they reveal about our brains. The 1996 paperback updates the story with molecular biology, gene cloning, and discovery of neurotransmitter receptors.

Evolving Brains By John Morgan Allman. Non-fiction. A distinguished contributor in his own right to brain research on vision, Allman brings a rare combination of neuroscience, evolutionary biology, and developmental biology to his work. *Evolving Brains* is a fascinating account of the uncanny, unconscious genius of evolution brilliantly improvising the brain in response to the needs of the gut, the blood, the hunt, and, always, the next generation.

Galen's Prophecy: Temperament in Human Nature By Jerome Kagan. Non-fiction. Psychologist Jerome Kagan takes a perceptive look at what research into infant and child development can teach us about human nature, in particular the biological influences on temperament.

How the Mind Works By Steven Pinker. Non-fiction. Pinker attempts to explain the brain's natural ability to perform feats that even the most sophisticated computer hardware would find impossible. He also explores how the mind thinks, reasons, falls in love, and develops family bonds.

Searching for Memory: The Brain, the Mind, and the Past By Daniel L. Schacter. Non-fiction. Schacter, chairman of psychology at Harvard, tells the story that brain research has found to explain the multiple, complex systems that underlie memory. We learn that with memory's power comes fragility, limitations seen not only in disease and aging but also in explosive issues such as "recovered memories" of child abuse that have put innocent teachers in prison.

The Broken Brain: The Biological Revolution in Psychiatry
By Nancy Andreasen. Non-fiction. Andreasen, a distinguished psychiatrist, introduces this book with chapters on the history of mental illness, the brain, the four major syndromes, diagnosis, treatment, and research. Many authors claim to write for laymen; Andreasen, a former English teacher, really does. Her subtext is that mental illness is a disease, no more shameful than cancer.

The Emotional Brain: The Mysterious Underpinnings of Emotional Life
By Joseph LeDoux. Non-fiction. The Emotional Brain reasons its way through questions about the nature of emotions, conservation of emotional systems across

species, conscious and unconscious emotional responses, and the relationship between feelings and emotions.

The Longevity Strategy: How to Live to 100 Using the Brain-Body Connection By David Mahoney and Richard Restak. Non-fiction. Mahoney, the business executive and philanthropist who was chairman of the Dana Alliance for Brain Initiatives, teamed up with the neurologist and neuropsychiatrist Restak for this road map to a healthy longevity. Includes 31 practical, research-based tactics for maintaining cognitive and emotional well-being, physical health, and financial stability through the life span.

The Man Who Mistook His Wife for a Hat and Other Clinical Tales By Oliver Sacks. Non-fiction. Patients with lesions and disorders have been a crucial window on the brain for neuroscientists. In this famous book. Sacks presents a series of such case studies, from Korsakov's syndrome, with its devastation of memory, to Tourette's syndrome, with its explosion of mental energy, in portraits that are profoundly revelatory and full of compassion for the afflicted individuals.

The Principles of Psychology By William James. Non-fiction. Stream of thought, consciousness of self, attention, conception, perception of time, memory: James analyzed, categorized, and conceptualized each aspect of mental life. Much remains valid— and not infrequently used as the starting point of discussions today—because James knew and honored the difference between observation and interpretation.

Why Zebras Don't Get Ulcers: The Acclaimed Guide to Stress, Stress-Related Diseases, and Coping, Third Edition By Robert M. Sapolsky. Non-fiction. Evolution of the fight-or-flight mechanism that, in a burst of physiological fireworks, can save a zebra from a lion, is often turned on—and left on—by the psychological and social stressors in our lives. Then the sympathetic nervous system's response to "danger" becomes the problem. Sapolsky explains all this, writing about glucocorticoids and insulin secretion with wit and charm.

Excerpted from: "Ourselves to Know" Books from Scientists of the Dana Alliance From Cerebrum: The Dana Forum on Brain Science, *Vol. 6, No. 2, Spring 2004 © Dana Press:*

Brave New Brain: Conquering Mental Illness in the Era of the Genome By Nancy Andreasen. Non-fiction. Describes progress over a decade or more in understanding the chief categories of mental illness (schizophrenia, dementia, mood disorders, and anxiety disorders), how treatments have changed, especially in light of understanding the genetics of illness, and what lies ahead. Includes a mini-tutorial on neuroscience and molecular genetics, a review of mental illnesses, and comments on what it all means in social and economic terms.

Looking for Spinoza: Joy, Sorrow, and the Feeling Brain By Antonio Damasio. Non-fiction. Examines the role in human existence of feelings and the emotions that underlie feelings (an important distinction for Damasio). Damasio shares his rediscovery of the 17th-century Dutch philosopher Benedict Spinoza whose *Ethics* defied the life-and death power of religion in his era in postulating the inseparability of mind and body. *Looking for Spinoza* is a complete work of philosophy as well as science, rooting a modern-day philosophy of human nature, the good life, and the just society in the discoveries of brain science.

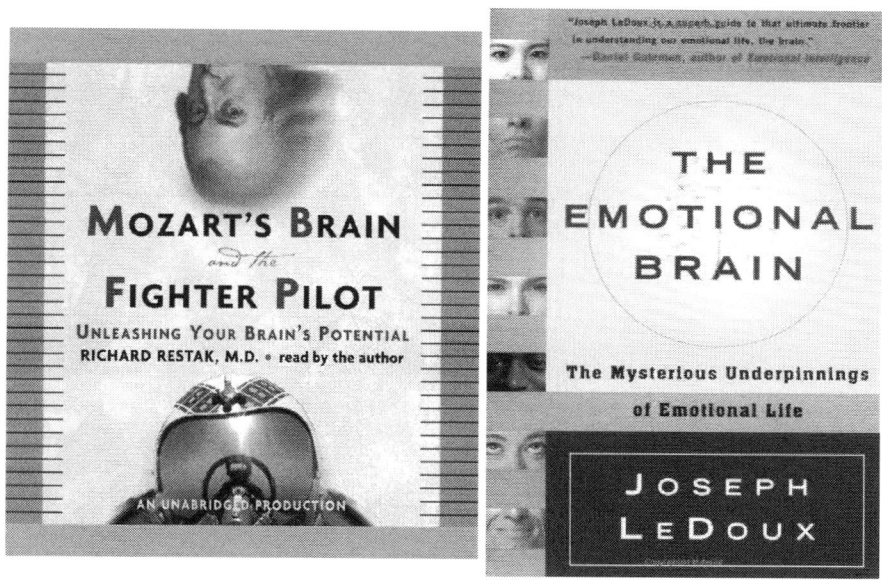

Memory and Emotion: The Making of Lasting Memories By James McGaugh. Non-fiction. McGaugh, professor of neurobiology and behavior at the University of California, Irvine, and pioneer in memory research, discusses the history of memory research, recent major discoveries, the molecular biological processes underlying formation of long-term memories, new ideas about post-traumatic stress disorder (PTSD), and more in an engaging and personable style.

Mozart's Brain and the Fighter Pilot: Unleashing Your Brain's Potential By Richard Restak. Non-fiction. Can we make ourselves smarter? Restak suggests in this book that a good grasp of how your brain works can even make you more intelligent. With more that 20 exercises intended to improve memory, creativity, and concentration by increasing neural linkages, this best-selling guide promises to enhance your cognitive capabilities now and into old age.

Mysteries of the Mind By Richard Restak. Non-fiction. *Mysteries of the Mind* takes readers on a tour of the brain, using drawings and illustrations to explore its structure and operation, particularly in sleep, memory, and emotion.

Overcoming Dyslexia: A New and Complete Science-Based Program for Reading Problems at Any Level By Sally Shaywitz. Non-fiction. A trusted authority on dyslexia for over 20 years, Shaywitz provides a comprehensive source of information and guidance, answering questions she has heard in years on the lecture circuit. All three sections of the book—on the nature of dyslexia, its diagnosis, and how to overcome it—are enlivened with personal stories from Shaywitz's work with students, teachers, and parents.

Imaginary Portraits By Walter Pater. Fiction. Rich interior descriptions of fictional characters, who functioned in part as psychological allegories. A favorite of teachers and students of literature, art history, and aesthetics.

Motherless Brooklyn By Jonathan Lethem. Fiction. Told in the voice of a character with Tourette syndrome. Set in modern-day Brooklyn, Lionel Essrog works to uncover who murdered his boss and learns that his tics actually make him a better detective.

The Diving Bell and the Butterfly By Jean-Dominique Bauby. Non-fiction. Describes what it's like to have one's intact subjectivity suddenly locked inside a body that will no longer move and can no longer speak. Bauby had an ultimately fatal stroke that resulted in a rare condition called "locked-in syndrome," leaving him able to move only his left eyelid.

The Man Who Tasted Shapes By Richard Cytowic. Non-fiction. An account of what it's like to live with synesthesia, a condition, according to the author, characterized by the prioritization of emotional knowledge over reason.

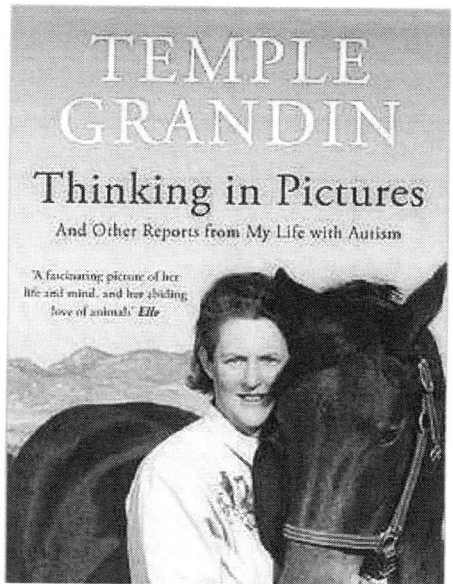

The Man With a Shattered World By A.R. Luria. Non-fiction. The story of a soldier, Zasetsky, who suffers brain damage and begins writing a journal to help put his thoughts, memories, and life back together. Told as a narrative in two voices, that of the patient as it comes through in excerpts from his journal, and that of his doctor, Luria himself, commenting on the patient's experiences and providing analytic context for making sense of them.

Thinking in Pictures, Expanded Edition: My Life With Autism
By Temple Grandin. Non-fiction. A woman with high-functioning autism points to ways which her highly visual, rational, and concrete form of thinking enables her to do things—such as designing more humane mechanisms for handling livestock—that so-called normal people cannot.

Excerpted from: Four Fictional Odysseys Through Life With a Disordered Brain By Todd E. Feinberg, M.D. From Cerebrum: The Dana Forum on Brain Science, *Vol. 7, No. 4, Fall 2005 © 2005 Dana Press:*

Born Twice: A Novel of Fatherhood By Giuseppe Pontiggia. Fiction. The story of how the members of a small family deal with each other, themselves, and the outside world in the face of their youngest child Paolo's cerebral palsy. Narrated by the father, the aptly named Frigerio, whose greatest concern seems to be how to love his son, how to be a good, loving and caring father despite his true feelings about Paolo's difficulties.

Lying Awake By Mark Salzman. Fiction. Sister John of the Cross has a religious epiphany for the first time in her life. At age 40, she has spent 20 years serving God dutifully in a monastery in the heart of Los Angeles, and now, finally, she feels fulfilled. Unfortunately, the seizures which caused the religious experience now threaten her health and Sister John's fellow nuns insist that she seek a neurological consultation.

Memory Book: A Benny Cooperman Mystery By Howard Engel. Fiction. The victim of an attempted murder, private eye Benny Cooperman is found in a dumpster beside a dead woman. Having sustained massive head injuries and unable to remember the past, make new memories, or even read his own handwriting, Benny has to solve a mystery and clear his name from his bed in a Toronto rehabilitation facility.

The Speed of Dark By Elizabeth Moon. Fiction. In a future America where science has all but eradicated autism, a stigma persists against those with even traces of the condition. Lou Arrendale's mild autism makes him a genius at his job doing pattern analysis for a large pharmaceutical company but renders him otherwise socially paralyzed. When a new genetic procedure that promises to reverse his autism becomes available, Lou has to decide if the change is worth it.

Excerpted from Brain Books for Budding Scientists—and All Children By Carolyn Phelan From Cerebrum: The Dana Forum on Brain Science, Vol. 4, No. 2, Summer 2002 © Dana Press:

101 Questions Your Brain Has Asked About Itself But Couldn't Answer...Until Now Faith Hickman Brynie. Non-fiction. Students' questions about the brain, paired with Brynie's answers, appear in seven chapters covering basic information, neurons, learning and memory, chemicals and drugs in the brain, damage and illness, left- and right-brain functions, and speech and the senses. Each chapter includes a related feature article on a topic such as brain imaging.

Head and Brain Injuries By Elaine Landau. Non-fiction. Landau surveys the most common forms of traumatic brain injuries, their causes and treatments, and how they change lives. In addition, she offers a brief historical survey of brain science and medical treatment.

Phineas Gage: A Gruesome but True Story About Brain Science By John Fleischman. Non-fiction. Here is the story of the 19th-century railway worker who accidentally drove an iron tamping rod into his skull. This case study marked the beginning of a fuller understanding of the brain. Readers new to Gage's tale will come away intrigued by the story, knowledgeable about the brain, and (even better) curious to find out more.

When the Brain Dies First By Margaret O. Hyde and John F. Setaro. The authors begin with a brief introduction to the healthy brain, then zero in on the many things that can go wrong. They discuss injuries to the head; encephalitis and Creutzfeldt-Jakob disease; dementia caused by Alzheimer's disease; and degenerative diseases such as multiple sclerosis, among many other topics.

The Curious Incident of the Dog in the Night-Time By Mark Haddon. Fiction. Haddon, who once worked with autistic children and currently teaches creative writing, leads readers into the chaos of autism. He creates a character of such empathy that many readers will feel for the first time what it is like to live with no filters to eliminate or order the millions of pieces of information that stimulate our senses every moment. The protagonist, an autistic teen, investigates the murder of a neighbor's dog, entering and coping with an overwhelming outside world

Appendix I

A Glossary of Key Brain Science Terms

In this collection of brain facts, you'll find terms and concepts key to understanding the brain and maintaining brain health. Typically, past study focused on structures in the brain. As you see them listed below, remember that structures, glands, hormones, regions, etc., work together in systems, which more accurately describes how our brains work. Adapted from the Dana Foundation, **www.dana.org**

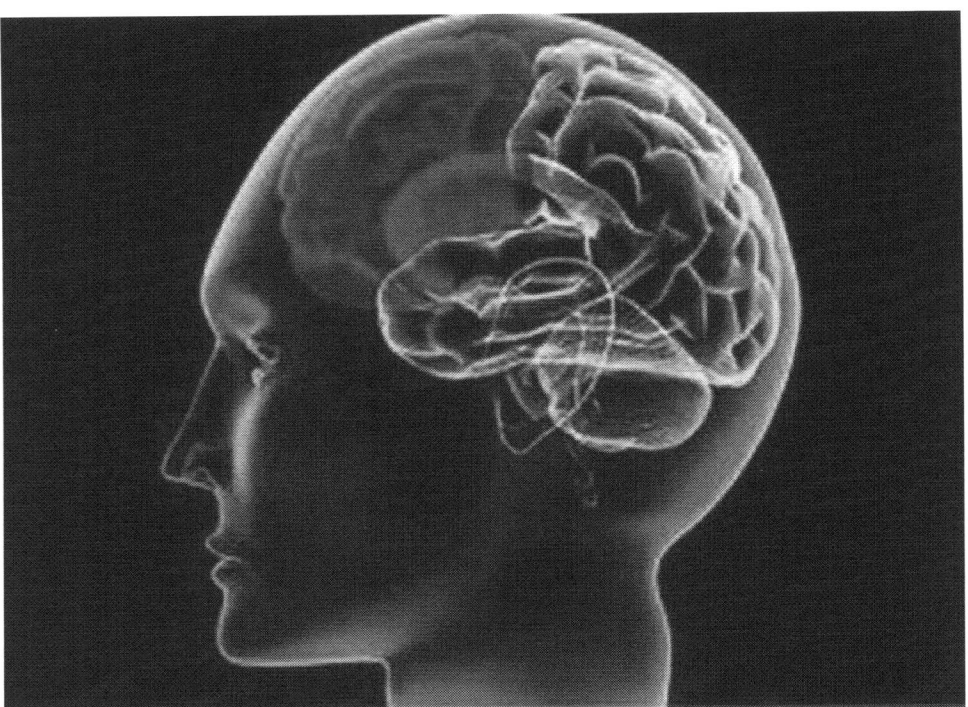

Wikimedia.com

GLOSSARY

A

adrenal glands: Located on top of each kidney, these two glands are involved in the body's response to stress and help regulate growth, blood glucose levels, and the body's metabolic rate. They receive signals from the brain and secrete several different hormones in response, including cortisol and adrenaline.

adrenaline: Also called epinephrine, this hormone is secreted by the adrenal glands in response to stress and other challenges to the body. The release of adrenaline causes a number of changes throughout the body, including the metabolism of carbohydrates to supply the body's energy demands.

allele: One of the variant forms of a gene at a particular location on a chromosome. Differing alleles produce variation in inherited characteristics such as hair color or blood type. A dominant allele is one whose physiological function—such as making hair blonde—is manifest even when only a single copy is present (among the two copies of each gene that everyone inherits from their parents). A recessive allele is one that manifests only when two copies are present.

amino acid: A type of small organic molecule. Amino acids have a variety of biological roles, but are best known as the "building blocks" of proteins.

amino acid neurotransmitters: The most prevalent neurotransmitters in the brain, these include glutamate and aspartate, which have excitatory actions, and glycine and gamma-amino butyric acid (GABA), which have inhibitory actions.

amygdala: Part of the brain's limbic system, this primitive brain structure lies deep in the center of the brain and is involved in emotional reactions, such as anger, as well as emotionally charged memories. It also influences behavior such as feeding, sexual interest, and the immediate "fight or flight" stress reaction that helps ensure that the body's needs are met.

amyloid-beta (Aβ) protein: A naturally occurring protein in brain cells. Large, abnormal clumps of this protein form the amyloid plaques that are the hallmark of Alzheimer's disease. Smaller groupings (oligomers) of Aβ seem more toxic to brain cells and are now thought by many researchers to be important initiators of the Alzheimer's disease process.

amyloid plaque: The sticky, abnormal accumulations of amyloid-beta protein aggregate around neurons and synapses in the memory and intellectual centers of the brain, in people with Alzheimer's. These are sometimes referred to as neuritic plaques or senile plaques. While amyloid plaques have long been considered markers of Alzheimer's, they are also found to some extent in many cognitively normal elderly people. Plaques' role in Alzheimer's neurodegeneration remains unclear.

animal model: A laboratory animal that—through changes in its diet, exposure to toxins, genetic changes, or other experimental manipulations—mimics specific signs or symptoms of a human disease. Many of the most promising advances in treating brain disorders have come from research on animal models.

astrocyte: A star-shaped glial cell that delivers "fuel" to the neurons from the blood, removes waste from the neuron, and otherwise modulates the activity of the neuron. Astrocytes also play critical roles in brain development and the creation of synapses.

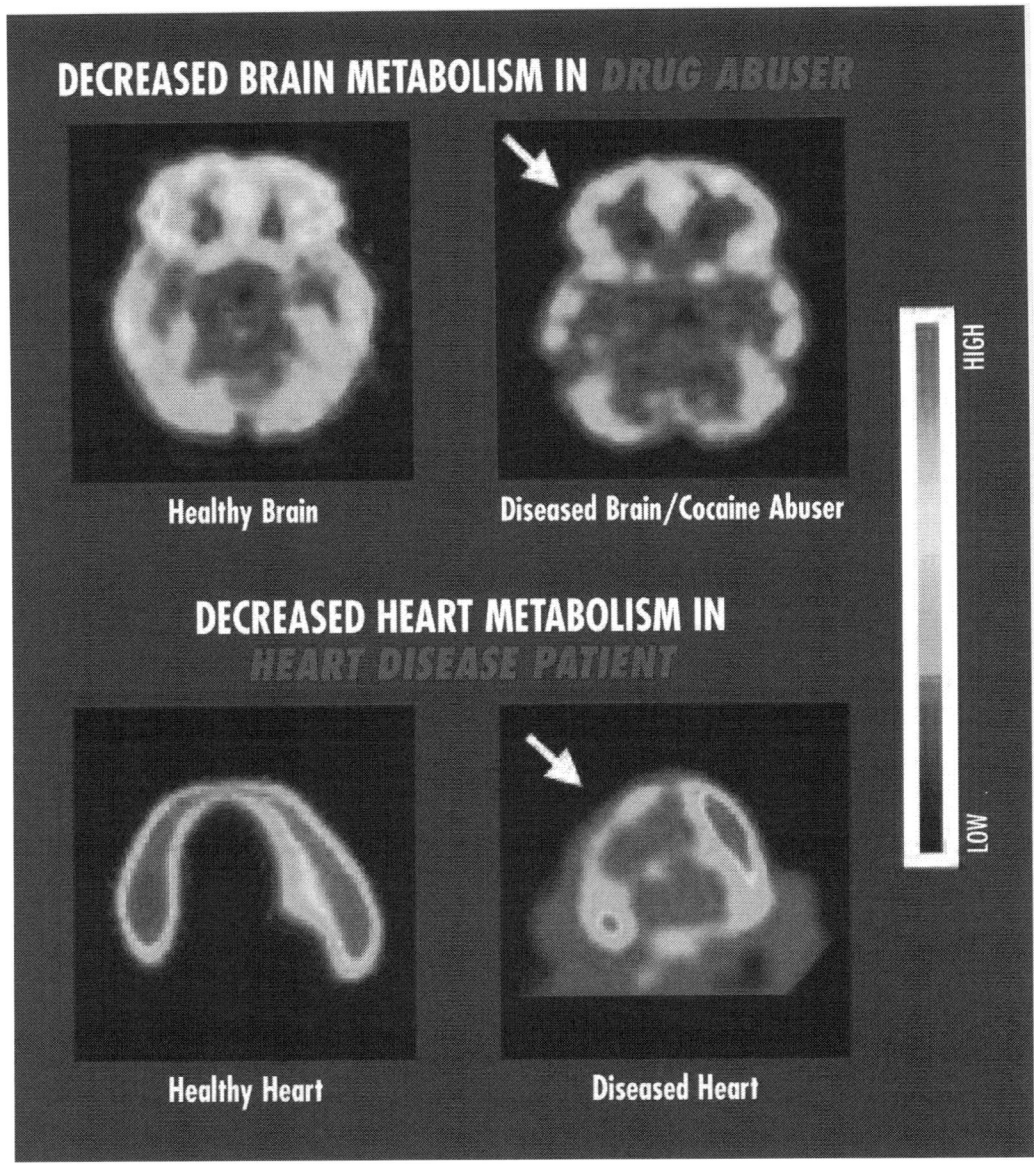

NIDA.gov

auditory cortex: Part of the brain's temporal lobe, this region is responsible for hearing. Nerve fibers extending from the inner ear carry nerve impulses generated by sounds into the auditory cortex for interpretation.

autonomic nervous system: Part of the central nervous system that controls functions of internal organs (e.g., blood pressure, respiration, intestinal function, urinary bladder control, perspiration, body temperature). Its actions are mainly involuntary, or "automatic."

axon: A long, single nerve fiber that transmits messages, via electrochemical impulses, from the body of the neuron to dendrites of other neurons, or directly to body tissues such as muscles.

B

basal ganglia: A group of structures below the cortex involved in motor, cognitive, and emotional functions.

basilar artery: Located at the base of the skull, the basilar artery is one of the major vascular components supplying oxygenated blood to the brain and nervous system.

biomarkers: A measurable physiological indicator of a biological state or condition. For example, amyloid plaques—as detected on amyloid PET scans, for example—are a biomarker of Alzheimer's disease. Biomarkers can be used for both diagnostic and therapeutic purposes.

brain-computer interface: A device or program that permits direct or indirect collaboration between the brain and a computer system. For example, a device that harnesses brain signals to control a screen cursor or prosthetic limb.

brain imaging: Refers to various techniques, such as magnetic resonance imaging (MRI), diffusion tensor imaging, and positron emission tomography (PET), that enable scientists to capture images of brain tissue and structure and to reveal what parts of the brain are associated with various behaviors or activities.

brain stem: A primitive part of the brain that connects the brain to the spinal cord. The brain stem controls functions basic to the survival in animal such as heart rate, breathing, digestive processes, and sleeping.

C

central nervous system: The brain and spinal cord constitute the central nervous system and are part of the broader nervous system, which also includes the peripheral nervous system.

central sulcus: The primary groove in the brain's cerebrum, which separates the frontal lobe in the front of the brain from the parietal and occipital lobes in the rear of the brain.

cerebellar artery: The major blood vessel providing oxygenated blood to the cerebellum.

cerebellum: A brain structure located at the top of the brain stem that coordinates the brain's instructions for skilled, repetitive movements and helps maintain balance and posture. Recent research also suggests the cerebellum may play a role, along with the cerebrum, in some emotional and cognitive processes.

cerebrum (also called cerebral cortex): The largest brain structure in humans, accounting for about two-thirds of the brain's mass and positioned over and around most other brain structures. The cerebrum is divided into left and right hemispheres, as well as specific areas called lobes that are associated with specialized functions.

chronic traumatic encephalopathy (CTE): Once known as dementia pugilistica and thought to be confined largely to former boxers, this progressive degenerative disease, with symptoms including impulsivity, memory problems, and depression, affects the brains of individuals who have suffered repeated concussions and traumatic brain injuries.

cognition: A general term that includes thinking, perceiving, recognizing, conceiving, judging, sensing, reasoning, and imagining. Also used as an adjective pertaining to cognition, as in "cognitive processes."

cognitive neuroscience: The field of study that investigates the biological mechanisms of cognition.

computational neuroscience: An interdisciplinary field of study that uses information processing properties and algorithms to further the study of brain function and behavior.

computed tomography (CT or CAT): An X-ray technique introduced in the early 1970s that enables scientists to take cross-sectional images of the body and brain. CT uses a series of X-ray beams passed through the body to collect information about tissue density, then applies sophisticated computer and mathematical formulas to create an anatomical image from the data.

consciousness: The state of being aware of one's feelings and what is happening around one; the totality of one's thoughts, feelings, and impressions. corpus callosum: The collection of nerve fibers connecting the two cerebral hemispheres.

cortex: The outer layer of the cerebrum. Sometimes referred to as the cerebral cortex.

cortisol: A steroid hormone produced by the adrenal glands that controls how the body uses fat, protein, carbohydrates, and minerals, and helps reduce inflammation. Cortisol is released in the body's stress response; scientists have found that prolonged exposure to cortisol has damaging effects on the brain.

CT scan (also called CAT scan): See computed tomography.

D

deep brain stimulation: A method of treating various neuropsychiatric and neurodegenerative disorders through small, controlled electric shocks administered from a special battery-operated neurostimulation implant. The implant, sometimes called a brain pacemaker, is placed within deep brain regions such as the globus pallidus or subthalamus.

default-mode network: The network indicates that the brain remains active even if not involved in a specific task. So whether asleep or daydreaming, the brain is in an active state.

delayed discounting: A common cognitive task used to measure impulsivity in individuals. The task measures an individual's preference for the immediate delivery of a small reward versus a larger reward delivered later.

dementia: General mental deterioration from a previously normal state of cognitive function due to disease or psychological factors. Alzheimer's disease is one form of dementia.

dendrites: Short nerve fibers that project from a nerve cell, generally receiving messages from the axons of other neurons and relaying them to the cell's nucleus.

depression: A mood or affective disorder characterized by sadness and lack of motivation. Depression has been linked to disruptions in one or more of the brain's neurotransmitter systems, including those related to serotonin and dopamine.

Clinical depression is a serious condition that can often be effectively treated with medications and/or behavioral therapy.

Diagnostic and Statistical Manual of Mental Disorders (DSM): The standard classification manual published by the American Psychiatric Association to be used by mental health professionals to diagnose and treat mental disorders.

DNA (deoxyribonucleic acid): The material from which the 46 chromosomes in each cell's nucleus is formed. DNA contains the codes for the body's approximately 30,000 genes, governing all aspects of cell growth and inheritance. DNA has a double-helix structure—two intertwined strands resembling a spiraling ladder.

dominant gene: A gene that almost always results in a specific physical characteristic, for example a disease, even though the patient's genome possesses only one copy. With a dominant gene, the chance of passing on the gene (and therefore the trait or disease) to children is 50-50 in each pregnancy.

dopamine: A neurotransmitter involved in motivation, learning, pleasure, the control of body movement, and other brain functions. Some addictive drugs greatly increase brain levels of dopamine, leading to a euphoric "high." Virtually all addictive substances, from nicotine to alcohol to heroin and crack cocaine, affect the dopamine system in one way or another.

double helix: The structural arrangement of DNA, which looks something like an immensely long ladder twisted into a helix, or coil. The sides of the "ladder" are formed by a backbone of sugar and phosphate molecules, and the "rungs" consist of nucleotide bases joined weakly in the middle by hydrogen bonds.

E

electroconvulsive therapy (ECT): A therapeutic treatment for depression and other mental illnesses that sends small electric currents over the scalp to trigger a brief seizure. It is one of the fastest ways known to reverse the symptoms of severely depressed individuals.

endocrine system: A body system composed of several different glands and organs that secrete hormones.

endorphins: Hormones produced by the brain, in response to pain or stress, to blunt the sensation of pain. Narcotic drugs, such as morphine, imitate the actions of the body's natural endorphins.

enzyme: A protein that facilitates a biochemical reaction. Organisms could not function if they had no enzymes.

epigenetics: A subset of genetics that focuses on phenotypic trait variations caused by specific environmental factors that influence where, when, and how a gene is expressed.

F

fissure: A groove or indentation observed in the brain. Another word for sulcus.

frontal lobe: The front of the brain's cerebrum, beneath the forehead. This area of the brain is associated with higher cognitive processes such as decision-making, reasoning, social cognition, and planning, as well as motor control.

functional magnetic resonance imaging (fMRI): A brain imaging technology, based on conventional MRI, that gathers information relating to short-term changes in oxygen uptake by neurons. It typically uses this information to depict the brain areas that become more active or less active—and presumably more or less involved—while a subject in the fMRI scanner performs a cognitive task.

G

GABA (gamma-aminobutyric acid): A neurotransmitter implicated in brain development, muscle control, and reduced stress response.

gene: The basic unit of inheritance. A gene is a distinct section of DNA in a cell's chromosome that encodes a specific working molecule—usually protein or RNA—with some role in brain or body function.

gene defects (genetic mutations) are thought to cause many disorders including brain disorders.

gene expression: The process by which a gene's nucleotide sequence is transcribed into the form of RNA—often as a prelude to being translated into a protein.

gene mapping: Determining the relative positions of genes on a chromosome and the distance between them.

genome: The complete genetic map for an organism. In humans, this includes about 30,000 genes, more than 15,000 of which relate to functions of the brain.

glia (or glial cells): The supporting cells of the central nervous system. Though probably not involved directly in the transmission of nerve signals, glial cells protect and nourish neurons.

glioma: A tumor that arises from the brain's glial tissue.

glucose: A natural sugar that is carried in the blood and is the principal source of energy for the cells of the brain and body. PET imaging techniques measure brain activity by detecting increases in the brain's metabolism of glucose during specific mental tasks.

gray matter: The parts of the brain and spinal cord made up primarily of groups of neuron cell bodies (as opposed to white matter, which is composed mainly of myelinated nerve fibers).

gyrus: The ridges on the brain's outer surface. Plural is gyri.

H

hemisphere: In brain science, refers to either half of the brain (left or right). The two hemispheres are separated by a deep groove, or fissure, down the center. Some major, specific brain functions are located in one or the other hemisphere.

hippocampus: A primitive brain structure, located deep in the brain, that is involved in memory and learning.

hormone: A chemical released by the body's endocrine glands (including the adrenal glands), as well as by some tissues. Hormones act on receptors in other parts of the body to influence body functions or behavior.

hypothalamus: A small structure located at the base of the brain, where signals from the brain and the body's hormonal system interact.

I

insula: Sometimes referred to as the insular cortex, this small region of the cerebrum is found deep within the lateral sulcus, and is believed to be involved in consciousness, emotion, and body homeostasis.

interneuronal: Between neurons, as in interneuronal communication.

ions: Atoms or small groups of atoms that carry a net electric charge, either positive or negative. When a nerve impulse is fired, ions flow through channels in the membrane of a nerve cell, abruptly changing the voltage across the membrane in that part of the cell. This sets off a chain reaction of similar voltage changes along the cell's axon to the synapse, where it causes the release of neurotransmitters into the synaptic cleft.

L

lesion: An injury, area of disease, or surgical incision to body tissue. Much of what has been learned about the functions of various brain structures or pathways has resulted from lesion studies, in which scientists observe the behavior of persons who have suffered injury to a distinct area of the brain or analyze the behavior resulting from a lesion made in the brain of a laboratory animal.

limbic system: A group of evolutionarily older brain structures that encircle the top of the brain stem. The limbic structures play complex roles in emotions, instincts, and behavioral drives.

M

MRI (magnetic resonance imaging): A non-invasive imaging technology, often used for brain imaging. An MRI scanner includes intensely powerful magnets, typically 10,000 to 40,000 times as strong as the Earth's magnetic field. These magnets, combined with coils that send electromagnetic pulses into the scanned tissue, induce radio-frequency signals from individual hydrogen atoms within the tissue. The scanner records and processes these signals to build up an image of the scanned tissue. MRI scans are able to depict high resolution images of the entire brain, allowing clinicians to determine if the brain tissue that is visualized is normal, abnormal, or damaged due to a neurological disorder or trauma. MRI technology has also been adapted to measure brain activity.

melatonin: A hormone that is secreted by the pineal gland in the brain in response to the daily light-dark cycle, influencing the body's sleep-wake cycle and possibly affecting sexual development.

memory: The encoding and storage of information, in a way that allows it to be retrieved later. In the brain, memory involves integrated systems of neurons in diverse brain areas, each of which handles individual memory-related tasks. Memory can be categorized into two distinct types, each with its own corresponding brain areas. Memory about people, places, and things—that one has experienced directly or otherwise learned about—is referred to as explicit or declarative memory and seems to be centered in the hippocampus and temporal lobe. Memory about motor skills and perceptual strategies is known as implicit, or procedural memory and seems to involve the cerebellum, the amygdala, and specific pathways related to the particular skill (e.g., riding a bicycle would involve the motor cortex).

metabolize: To break down or build up biochemical elements in the body, effecting a change in body tissue. Brain cells metabolize glucose, a blood sugar, to derive energy for transmitting nerve impulses.

microbiota: The community of various microorganisms found in the digestive tract. Scientists are now learning that microbes found in the microbiota can influence brain development and behavior.

microglia: A small, specialized glial cell that operates as the first line of immune defense in the central nervous system.

minimally conscious state: A disorder of consciousness, often caused by stroke, head injury, or loss of blood flow to the brain, in which an individual maintains partial conscious awareness.

molecular biology: The study of the structure and function of cells at the molecular level and how these molecules influence behavior and disease processes. Molecular biology emerged as a scientific discipline only in the 1970s, with advances in laboratory technologies for isolating and characterizing DNA, RNA, proteins, and other types of biological molecule.

motor cortex: The part of the brain's cerebrum, just to the front of the central sulcus in the frontal lobe, that is involved in movement and muscle coordination. Scientists have identified specific spots in the motor cortex that control movement in specific parts of the body, the so-called "motor map."

MRI: See magnetic resonance imaging and/or functional magnetic resonance imaging.

mutation: A permanent structural alteration to DNA that alters its previous nucleotide sequence. In most cases, DNA changes either have no effect or cause harm, but occasionally a mutation improves an organism's chance of surviving and procreating.

myelin: The fatty substance that sheathes most nerve cell axons, helping to insulate and protect the nerve fiber and effectively speeding up the transmission of nerve impulses.

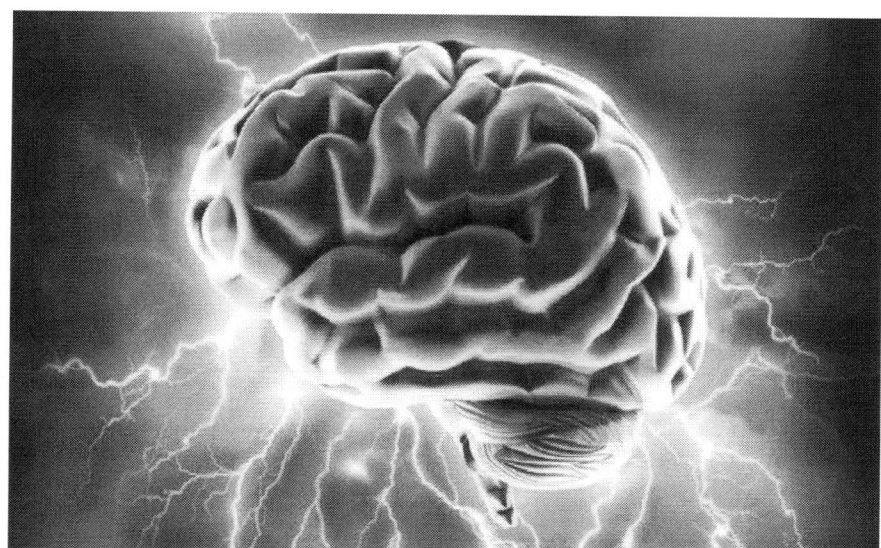
Wikimedia.com

N

narcotic: A synthetic chemical compound that mimics the action of the body's natural endorphins—hormones secreted to counteract pain. Narcotic drugs have a valid and useful role in the management of pain but may lead to physical dependence in susceptible individuals if used for long periods.

neurodegenerative diseases: Diseases characterized by the progressive deterioration and death of nerve cells (neurodegeneration), typically originating in one area of the brain and spreading to other connected areas. Neurodegenerative diseases include amyotrophic lateral sclerosis (also known as Lou Gehrig's disease), Huntington's disease, Alzheimer's disease, frontotemporal degeneration, and Parkinson's disease.

neuroethics: An interdisciplinary field of study that addresses the ethical issues of our increased ability to understand and change the brain. Privacy, life extension, cloning, and many other issues are included in this ongoing social-scientific debate.

neurogenesis: The production of new, maturing neurons by neural stem and progenitor cells. Rapid and widespread neurogenesis obviously occurs in the fetal brain in humans and other animals. Neuroscientists long believed that neurogenesis essentially does not occur in the adult human brain. However, over the past two decades, research has shown that it does in fact occur in the dentate gyrus of the hippocampus and possibly other brain regions. This "adult neurogenesis" appears to be vital for normal learning and memory, and may help protect the brain against stress and depression. Neural stem cells, which can produce new, "young" neurons and glial cells, also may be used widely someday to treat brain disorders, particularly neurodegenerative diseases that otherwise deplete the neuronal population.

neuron: Nerve cell. The basic unit of the central nervous system, the neuron is responsible for the transmission of nerve impulses. Unlike any other cell in the body, a neuron consists of a central cell body as well as several threadlike "arms" called axons and dendrites, which transmit nerve impulses. Scientists estimate that there are approximately 100 billion neurons in the brain.

neuroscience: The study of brains and nervous systems, including their structure, function, and disorders. Neuroscience as an organized discipline gained great prominence in the latter part of the last century.

neurotransmitter: A chemical that acts as a messenger between neurons and is released into the synaptic cleft when a nerve impulse reaches the end of an axon. Several dozen neurotransmitters have been identified in the brain so far, each with specific, often complex roles in brain function and human behavior.

nurture: A popular term for the influence of environmental factors on human development such as the experiences one is exposed to in early life. The term is often used in the context of "nature versus nurture," which relates to the interplay of "nature" (genetic or inherited, predetermined influences) and environmental, or experiential, forces.

O

occipital lobe: A part of the brain's cerebrum, located at the rear of the brain, above the cerebellum. The occipital lobe is primarily concerned with vision and encompasses the visual cortex.

olfactory: Pertaining to the sense of smell. When stimulated by an odor, olfactory receptor cells in the nose send nerve impulses to the brain's olfactory bulbs, which in turn transmit the impulses to olfactory centers in the brain for interpretation.

opiate: A synthetic (e.g., Demerol, Fentanyl) or plant-derived (e.g., opium, heroin, morphine) compound that binds and activates opioid receptors on certain neurons. Opiates typically but not always have pain-relieving, anxiety-reducing, and even euphoria-inducing effects, and are generally considered addictive.

opioid: An artificially derived drug or chemical that acts on the nervous system in a similar manner to opiates, influencing the "pleasure pathways" of the dopamine system by locking on to specialized opioid receptors in certain neurons.

opioid receptors (e.g., mu, delta, kappa): A class of receptors found on neurons in the brain, spinal cord, and digestive tract. Opioid receptors are involved in numerous functions, including pain control, mood, digestion, and breathing.

oxytocin: Sometimes referred to as the "cuddle chemical," this hormone can work as a neurotransmitter in the brain and has been linked to social attachment and parental care.

P

pain receptors: Specialized nerve fibers in the skin and on the surfaces of internal organs, which detect painful stimuli and send signals to the brain.

parietal lobe: The area of the brain's cerebrum located just behind the central sulcus. It is concerned primarily with the reception and processing of sensory information from the body and is also involved in map interpretation and spatial orientation (recognizing one's position in space vis-a-vis other objects or places).

peripheral nervous system: The nervous system outside the brain and spinal cord.

persistent vegetative state: A disorder of consciousness, often following severe brain trauma, in which an individual has not even minimal conscious awareness. The condition can be transient, marking a stage in recovery, or permanent.

PET (positron emission tomography). An imaging technique, often used in brain imaging. For a PET scan of the brain, a radioactive "marker" that emits, or releases, positrons (parts of an atom that release gamma radiation) is injected into the bloodstream. Detectors outside of the head can sense these "positron emissions," which are then reconstructed using sophisticated computer programs to create "tomographs," or computer images. Since blood flow and metabolism increase in brain regions at work, those areas have higher concentrations of the marker, and researchers are able to see which brain regions are activated during certain tasks or exposure to sensory stimuli. Ligands can be added to a PET scan in order to detect pathological entities such as amyloid or tau deposits.

pharmacotherapy: The use of pharmaceutical drugs for therapeutic purposes.

pituitary gland: An endocrine organ at the base of the brain that is closely linked with the hypothalamus. The pituitary gland is composed of two lobes and secretes a number of hormones that regulate the activity of the other endocrine organs in the body.

plasticity: In neuroscience, refers to the brain's capacity to change and adapt in response to developmental forces, learning processes, injury, or aging.

postsynaptic cell: The neuron on the receiving end of a nerve impulse transmitted from another neuron.

prefrontal cortex: The area of the cerebrum located in the forward part of the frontal lobe, which mediates many of the higher cognitive processes such as planning, reasoning, and "social cognition"—a complex skill involving the ability to assess social situations in light of previous experience and personal knowledge, and interact appropriately with others. The prefrontal cortex is thought to be the most recently evolved area of the brain.

premotor cortex: The area of the cerebrum located between the prefrontal cortex and the motor cortex, in the frontal lobe. It is involved in the planning and execution of movements.

presynaptic cell: In synaptic transmission, the neuron that sends a nerve impulse across the synaptic cleft to another neuron.

psychiatry: A medical specialty dealing with the diagnosis and treatment of mental disorders. (Contrast with psychology).

psychoactive drug: A broad term for any drug that acts on the brain and noticeably alters one's mental state such as by elevating mood or alertness, or reducing inhibitions. Psychoactive pharmaceuticals can help control the symptoms of some neurological and psychiatric disorders. Many "recreational drugs" are also psychoactive drugs.

psychological dependence: In the science of addiction, refers to the mood- and motivation-related factors that sustain addiction (such as craving a cigarette after a meal), as opposed to the "physical dependence" that manifests when a person attempts to kick the habit (e.g., tremors, racing pulse). Brain scientists now understand that psychological factors are central to addictive disorders and are often the most difficult to treat. (Also see dependence.)

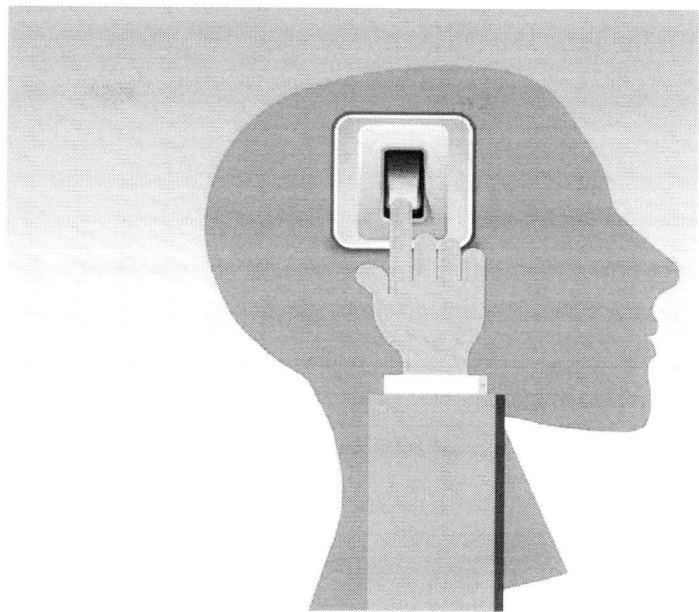
Pixabay

psychology: An academic or scientific field of study concerned with the behavior of humans and animals and related mental processes. (Contrast with psychiatry.)

PTSD (post-traumatic stress disorder): A mental disorder that develops in response to a traumatic event such as combat, sexual assault, terrorism, or abuse. Symptoms can include mood disturbances, hyperarousal, memory flashbacks, sleep problems, anxiety, and depression.

R

receptors: Molecules on the surfaces of neurons whose structures precisely match those of chemical messengers (such as neurotransmitters or hormones) released during synaptic transmission. The chemicals attach themselves to the receptors, in lock-and-key fashion, to activate the receiving cell structure (usually a dendrite or cell body).

recessive: A genetic trait or disease that appears only in patients who have received two copies of a mutant gene, one from each parent.

resting state: The state of the brain when it is not consciously engaged in an explicit task. Brain imaging techniques such as fMRI can be used to measure the residual activity that occurs in this state.

reward/reinforcement brain network: Also known as the mesolimbic circuit, this important network of brain regions is implicated in risk and reward processing, as well as learning. It primarily uses dopamine for signaling.

reuptake: A process by which released neurotransmitters are absorbed for subsequent re-use.

RNA (ribonucleic acid): A chemical similar to a single strand of DNA. RNA delivers DNA's genetic message to the cytoplasm of a cell, where proteins are made.

S

senses: The physiological inputs that provide critical information for perception and behavior from the outside world. The five classic senses are: sight, hearing, taste, touch, and smell.

serotonin: A neurotransmitter believed to play many roles, including, but not limited to, temperature regulation, sensory perception, and the onset of sleep. Neurons using serotonin as a transmitter are found in the brain and in the gut. A number of antidepressant drugs are targeted to brain serotonin systems.

social neuroscience: The field of study investigating the biological systems underlying social processes and behavior.

spinal cord: The "other half" of the central nervous system (with the brain). The spinal cord is a cable that descends from the brain stem to the lower back. It consists of an inner core of gray matter surrounded by white matter.

stem cells: Undifferentiated cells that can grow into heart cells, kidney cells, or other cells of the body. Originally thought to be found only in embryos, stem cells in the brain have unexpectedly been discovered in adults. Researchers have shown on research animals that stem cells can be transplanted into various regions of the brain, where they develop into both neurons and glia.

sulcus: The shallower grooves on the brain's cerebrum (deeper grooves are called fissures). Plural is sulci.

synapse: The junction where an axon approaches another neuron or its extension (a dendrite); the point at which nerve-to-nerve communication occurs. Nerve impulses traveling down the axon reach the synapse and release neurotransmitters into the synaptic cleft, the tiny gap between neurons.

synaptic transmission: The process of cell-to-cell communication in the central nervous system, whereby one neuron sends a chemical signal across the synaptic cleft to another neuron.

T

tau protein: A type of protein abundantly found in neurons. When this protein is not adequately cleared from the brain, it can form tangles that are a key pathology of several neurodegenerative disorders including frontotemporal degeneration, CTE, and Alzheimer's disease.

telomere: The protective cap found at the end of a chromosome. Research studies suggest that these caps may be shortened in neurodegenerative disorders.

temporal lobes: The parts of the cerebrum that are located on either side of the head, roughly beneath the temples in humans. These areas are involved in hearing, language, memory storage, and emotion.

thalamus: A brain structure located at the top of the brain stem, the thalamus acts as a two-way relay station, sorting, processing, and directing signals from the spinal cord and mid-brain structures to the cerebrum, and from the cerebrum down.

V

vagus nerve stimulation: A treatment for epilepsy that involves a small implant that electrically stimulates the vagus nerve, which runs from the brainstem to the abdomen.

vestibular: Refers to the sense of balance. Many people with hearing loss also have some degree of balance difficulties, since the vestibular (or balance) system and the auditory (or hearing) systems are so closely related.

visual cortex: The area of the cerebrum that is specialized for vision. It lies primarily in the occipital lobe at the rear of the brain, and is connected to the eyes by the optic nerves.

W

white matter: Brain or spinal cord tissue consisting primarily of the myelin-covered axons that extend from nerve cell bodies in the gray matter of the central nervous system.

Sources from the original Glossary, published in 2006, are included here, in addition to new updated sources reflecting the continuing evolution of neuroscience research.

- The Human Connectome Project, www.humanconnectomeproject.org. • The Mayo Clinic, www.mayoclinic.org. • Mind Over Matter Teacher's Guide, Introduction and Background, National Institute of Drug Abuse (NIDA), teens.drugabuse.gov/teachers/mind-over-matter/teachers-guide. • National Human Genome Research Institute (NIH), www.nhgri.nih.gov. • National Institutes of Health, NIH.gov. • Society for Neuroscience, BrainFacts.org. • Marcus S, ed., Neuroethics: Mapping the Field. Dana Press; Washington, DC (2002). • Clayman C, ed. The Human Body: An Illustrated Guide to Its Structure, Function, and Disorders. Dorling Kindersley; New York (1995). • Posner MI, Raichle ME. Images of Mind. Scientific American Library; New York (1994). • Blazing a Genetic Trail, Howard Hughes Medical Institute, www.hhmi.org (1991). • Webster's New World Dictionary, 3rd College Edition. Simon & Schuster; New York (1991). • Stedman's Medical Dictionary, 24th Edition. Williams & Wilkins; Baltimore (1982).

Scientific Advisor: Jordan Grafman, Ph.D. Published 2016

Appendix II

Maps of the Brain

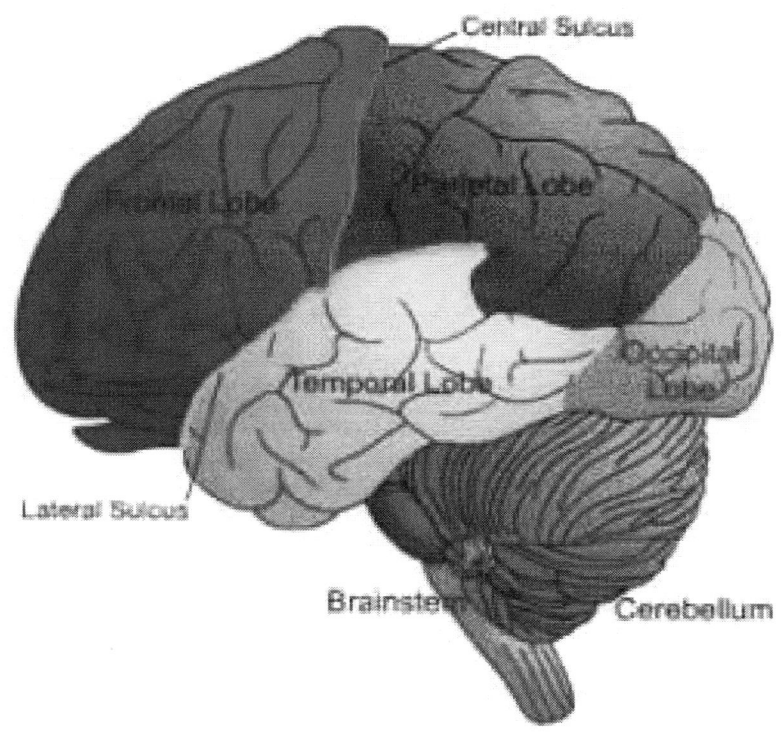

NIDA
Brain research has allowed us to learn the roles and functions of major brain centers. The brain stem and cerebellum, primitive brain structures, play key roles in memory and learning.

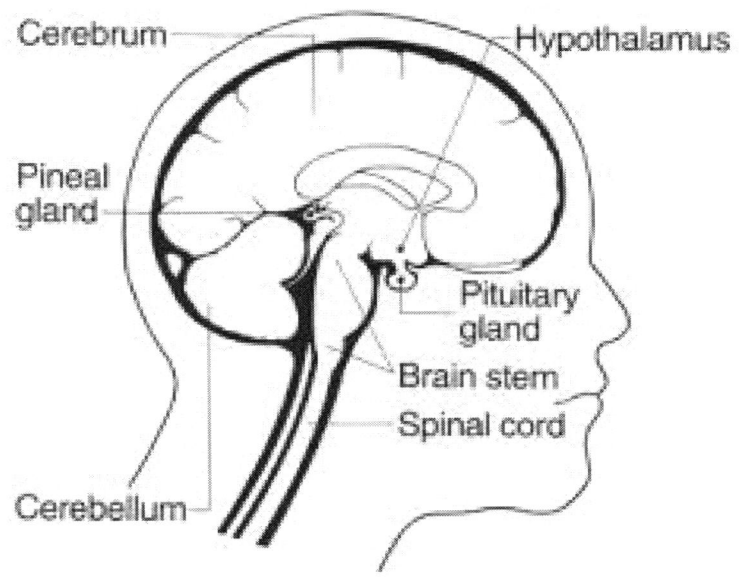

NIDA
The pineal gland reacts to light and produces melatonin, key to our sleep/wake cycles.

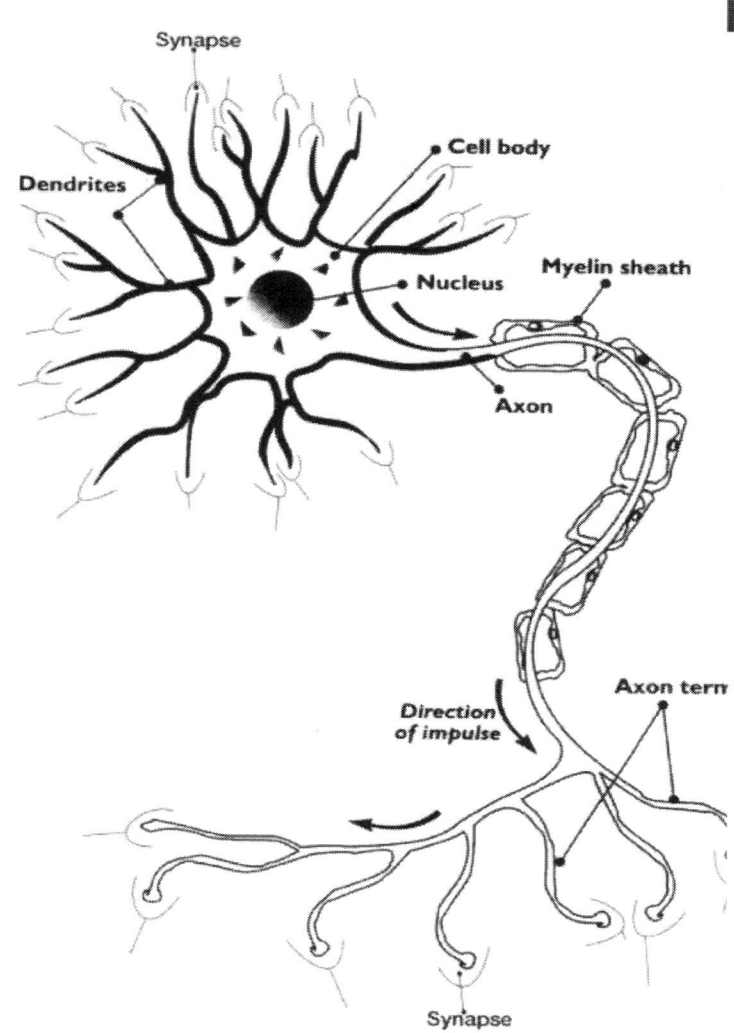

HHMI: Blazing a Genetic Trail
Neurons, or brain cells, process messages at an astonishing 268 miles per hour.

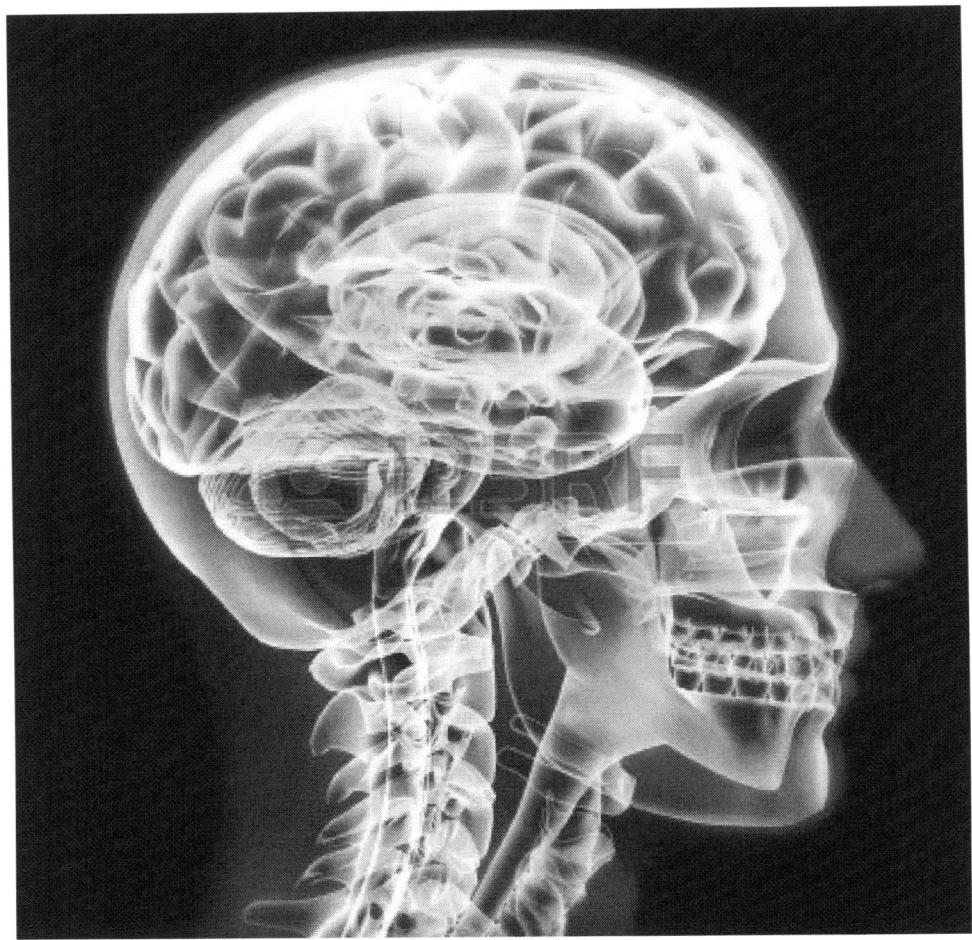

Wikimedia Commons
The brain and the spinal cord combine to form the central nervous system.

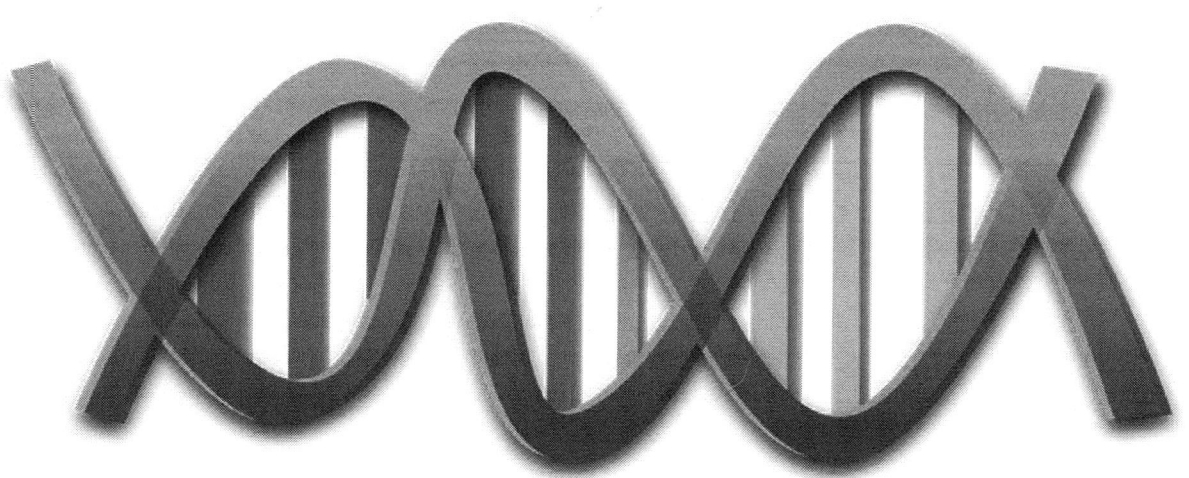

HHMI: Blazing a Genetic Trail
The double helix chain of DNA is found in each cell of the body, within the nucleus. Our genes govern all aspects of cell/body growth and inheritance.

Appendix III
Learn More About It
Resources on the Brain

Public Domain Pictures

GENERAL INFORMATION/SUPPORT

American Academy of Neurology: www.aan.com
American Association of Suicidology: www.suicidology.org
American Psychiatric Association: www.psych.org
American Psychological Association: www.apa.org
Brain and Behavior Researh Foundation (formerly NARSAD) www.bbrfoundation.org
Easter Seals: www.easter-seals.org
Family Caregiver Alliance: www.caregiver.org
Federation of Families for Children's Mental Health: www.ffcmh.org Genetic Alliance: www.geneticalliance.org
Mental Health America: www.nmha.org
National Alliance on Mental Illness (NAMI): www.nami.org
National Coalition of Creative Arts Therapies: www.nccata.org

National Mental Health Association: www.nmha.org
National Organization for Rare Disorders: www.rarediseases.org

GOVERNMENT RESOURCES
Centers for Disease Control and Prevention: www.cdc.gov
National Cancer Institue (NIH): www.cancer.gov
National Center for Complementary and Alternative Medicine: Information Clearinghouse: www.nccam.nih.gov
National Eye Institute: www.nei.nih.gov
National Heart, Lung and Blood Institute: www.nhlbi.nih.gov
National Institute on Alcohol Abuse and Alcoholism: www.niaaa.nih.gov
National Institute of Allergy and Infectious Diseases: www.niaid.nih.gov
Eunice Kennedy Shriver National Institute of Child Health and Human Development: www.nichd.nih.gov
National Institute on Deafness and Other Communication Disorders: www.nidcd.nih.gov
National Institute on Drug Abuse (NIDA): www.nida.nih.gov
National Institute of Mental Health: www.nimh.nih.gov
National Institute of Neurological Disorders and Stroke: Office of Communications and Public Liaison: www.ninds.nih.gov
Office of Rare Diseases at NIH: www.rarediseases.info.nih.gov

LGBT/Q General
CDC National Prevention Information Network: www.cdcnpin.org
Gay Men's Health Crisis (GMHC): www.gmhc.org

ALCOHOL AND DRUG ABUSE
Al-Anon Family Groups Headquarters, Inc.: www.al-anon.alateen.org
Alcoholics Anonymous: AA World Services, Inc.: www.aa.org
National Clearinghouse for Alcohol & Drug Abuse Information: www.health.org
National Council on Alcoholism and Drug Dependence (NCADD): www.ncadd.org
Recovery International Inc.: www.lowselfhelpsystems.org

ANOREXIA/BULIMIA (See: Eating Disorders)

ANXIETY DISORDERS
Anxiety Disorders Association of America: www.adaa.org

Freedom From Fear: www.freedomfromfear.org
National Anxiety Foundation: www.lexington-on-line.com/naf.html

APHASIA
National Aphasia Association: www.aphasia.org

ATAXIA AND ATAXIA-TELANGIECTASIA
A-T Children's Project: www.atcp.org
National Ataxia Foundation: www.ataxia.org

ATTENTION DEFICIT/HYPERACTIVITY DISORDER (ADHD) (See also: Learning Disabilities)
Children and Adults With Attention Deficit/Hyperactivity Disorder (CHADD): www.chadd.org

AUTISM
Autism Genetic Resource Exchange: www.agre.org
Autism Society of America: www.autism-society.org
Autism Speaks: www.autismspeaks.org

AUTOIMMUNE DISEASES (See: Neuroimmunological Disorders)

BEHAVIOR THERAPY
Association for Advancement of Behavior Therapy (AABT): www.aabt.org

BEHCET'S DISEASE American Behcet's Disease Association: www.behcets.com

BENIGN ESSENTIAL BLEPHAROSPASM
Benign Essential Blepharospasm Research Foundation: www.blepharospasm.org

BIRTH DEFECTS
Birth Defect Research for Children: www.birthdefects.org
March of Dimes Birth Defects Foundation: www.marchofdimes.com National Center on Birth Defects and Developmental Disabilities: www.cdc.gov

BLINDNESS/VISION IMPAIRMENT

Helen Keller National Center for Deaf/Blind Youth and Adults: www.hknc.org
Prevent Blindness America: www.preventblindness.org
Research to Prevent Blindness: www.rpbusa.org

BORDERLINE PERSONALITY DISORDER
Treatment and Research Advancements National Association for Personality Disorders: www.tara4bpd.org

BRAIN INJURY/PREVENTION
Brain Injury Association of America/Family HelpLine: www.biausa.org

Head Injury Hotline: www.headinjury.com
Think First Foundation: www.thinkfirst.org

BRAIN TUMOR (See also: Pediatric Brain Tumor; Pituitary Disorders)
American Brain Tumor Association: www.abta.org
The Brain Tumor Society: www.tbts.org
Dana-Farber Cancer Institute: www.dana-farber.org
The Healing Exchange BRAIN TRUST: www.braintrust.org
National Brain Tumor Foundation: www.braintumor.org

CEREBRAL PALSY
UCP National (United Cerebral Palsy)/United Cerebral Palsy Research and Educational Foundation: www.ucp.org

CHARCOT-MARIE-TOOTH DISEASE
Charcot-Marie-Tooth Association: www.cmtausa.org

CHIARI MALFORMATION (See: Spina Bifida; Syringomyelia)

COMA (See also: Brain Injury/Prevention)
Coma Recovery Association: www.comarecovery.org

CONCUSSION (See: Brain Injury/Prevention)

DEAFNESS/HEARING LOSS
Alexander Graham Bell Association for the Deaf and Hard of Hearing: www.agbell.org

American Society for Deaf Children: www.deafchildren.org
Better Hearing Institute: www.betterhearing.org
National Cued Speech Association: www.cuedspeech.org
Self Help for Hard of Hearing People: www.shhh.org

DEJERINE-SOTTAS DISEASE
(See: Muscular Dystrophy)

DEPRESSION/MANIC DEPRESSION
Depression and Bi-Polar Support Alliance www.dbsalliance.org

Depression and Related Affective Disorders Association (DRADA): www.drada.org

DISABILITY AND REHABILITATION (See also: Spinal Cord Injury)
National Information Center for Children and Youth With Disabilities: www.nichcy.org

DIZZINESS (See: Vestibular Disorders)

DOWN SYNDROME
National Down Syndrome Society: www.ndss.org

DRUG ABUSE (See also: Alcohol and Drug Abuse)
Do It Now Foundation: www.doitnow.org
National Families in Action: www.nationalfamilies.org

DYSAUTONOMIA
The Dysautonomia Foundation: www.familialdysautonomia.org

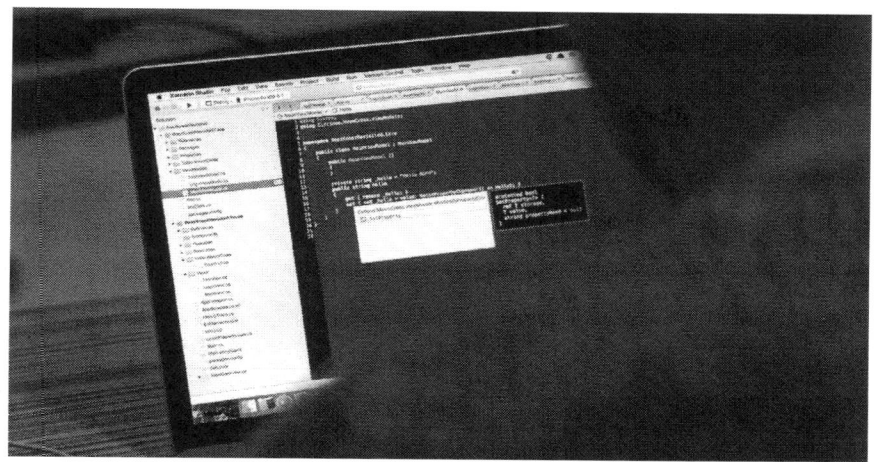
Pixabay.com

DYSLEXIA (See also: Learning Disabilities)
The International Dyslexia Association: www.interdys.org

DYSTONIA (See also: Tardive Dyskinesia/Tardive Dystonia) Dystonia Medical Research Foundation: www.dystonia-foundation.org

EATING DISORDERS
National Association of Anorexia Nervosa and Associated Disorders: www.anad.org

National Eating Disorders Association (ANRED)
www.nationaleatingdisorders.org

EPILEPSY
Epilepsy Foundation: www.efa.org

ESSENTIAL TREMOR/FAMILIAL TREMOR
International Essential Tremor Foundation: www.essentialtremor.org

FETAL ALCOHOL SYNDROME
National Organization on Fetal Alcohol Syndrome: www.nofas.org

FRAGILE X SYNDROME FRAXA
Research Foundation: www.fraxa.org

FRIEDREICH'S ATAXIA (See: Muscular Dystrophy)

GAUCHER DISEASE
National Gaucher Foundation: www.gaucherdisease.org

GUILLAIN-BARRE SYNDROME
Guillain-Barre Syndrome/CIPD Foundation International: www.gbs-CIPD.org

HEADACHE
American Headache Society: www.achenet.org
Association for Applied Psychophysiology and Biofeedback: www.aapb.org
National Headache Foundation: www.headaches.org

HEAD INJURY/TRAUMA (See: Brain Injury/Prevention; Coma) j

HUNTINGTON'S DISEASE
Hereditary Disease Foundation: www.hdfoundation.org Huntington's Disease Society of America: www.hdsa.org

HYDROCEPHALUS
Hydrocephalus Association: www.hydroassoc.org
National Hydrocephalus Foundation: www.nhfonline.org

JOSEPH DISEASES (See: Rare Disorders)

JOUBERT SYNDROME
Joubert Syndrome & Related Disorders Foundation: www.jsrcd.com

LEARNING DISABILITIES
Learning Disabilities Association of America: www.ldaamerica.org
National Center for Learning Disabilities: www.ncld.org

LEIGH'S DISEASE (See: Rare Disorders)

LEUKODYSTROPHY
United Leukodystrophy Foundation: www.ulf.org

LOWE SYNDROME
Lowe Syndrome Association: www.lowesyndrome.org

LUPUS
Lupus Foundation of America: www.lupus.org

MACHADO-JOSEPH DISEASES (See: Rare Disorders)

MEIGE SYNDROME (See: Benign Essential Blepharospasm)

MENTAL RETARDATION
The Arc of the United States: www.thearc.org

MOEBIUS SYNDROME (See: Rare Disorders)

MULTIPLE SCLEROSIS
Multiple Sclerosis Association of America: www.msassociation.org
Multiple Sclerosis Foundation: www.msfocus.org
National Multiple Sclerosis Society: www.nationalmssociety.org

MUSCULAR DYSTROPHY
Muscular Dystrophy Association: www.mdausa.org

MYASTHENIA GRAVIS (See also: Muscular Dystrophy)
Myasthenia Gravis Foundation of America: www.myasthenia.org

MYOSITIS (See also: Muscular Dystrophy)
Myositis Association of America: www.myositis.org

NARCOLEPSY (See also: Sleep Disorders)
Narcolepsy Network: www.narcolepsynetwork.org

NEIMANN-PICK DISEASE (See: Rare Disorders) j

NEUROFIBROMATOSIS
Children's Tumor Foundation: www.ctf.org
Neurofibromatosis Inc.: www.nfinc.org

NEUROIMMUNOLOGICAL DISORDERS (See also: Multiple Sclerosis; Myasthenia Gravis)
American Autoimmune Related Diseases Association: www.aarda.org

NEUROMUSCULAR DISEASES (See also: Polio)
Post-Polio Health International: www.post-polio.org

NEUROVASCULAR DISEASES (See: Stroke; Epilepsy; Brain Tumor)

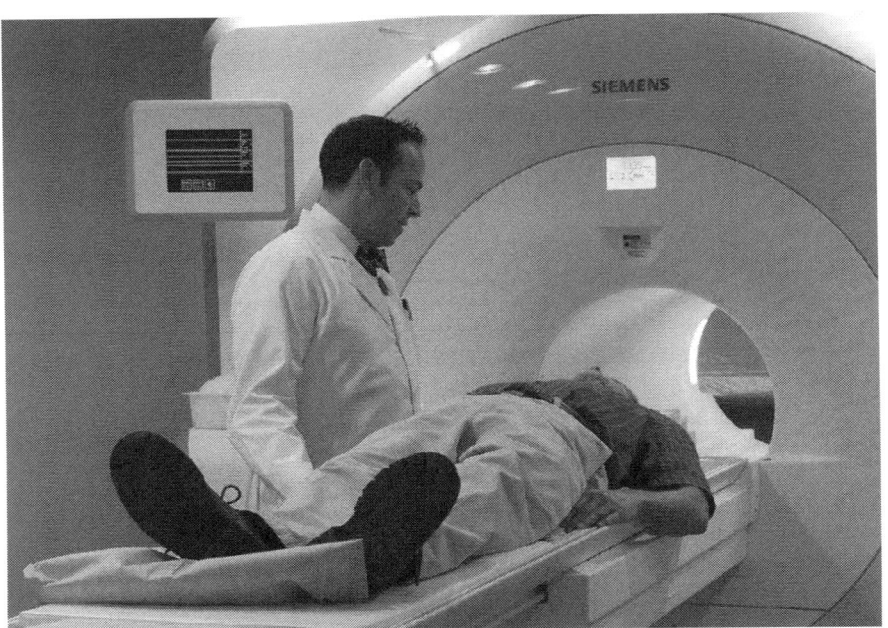

Flickr.com

OBSESSIVE-COMPULSIVE DISORDER
(OCD) International Obsessive-Compulsive Foundation:
www.ocfoundation.org
Trichotillomania Learning Center: www.trich.org

PAIN (CHRONIC)
American Chronic Pain Association: www.theacpa.org
American Pain Foundation: www.painfoundation.org

PANIC DISORDERS (See: Anxiety Disorders) j

PARALYSIS (See: Spinal Cord Injury; Disability and Rehabilitation)

PARKINSON'S DISEASE
American Parkinson's Disease Association: www.apdaparkinson.com
Parkinson Action Network: www.pdf.org

Parkinson's Action Network: www.parkinsonsaction.org
The Parkinson's Institute: www.pi.org

PEDIATRIC BRAIN TUMOR
Brain Tumor Foundation for Children: www.braintumorkids.org
The Childhood Brain Tumor Foundation: www.childhoodbraintumor.org
Children's Brain Tumor Foundation: www.cbtf.org
Dana-Farber Cancer Institute: www.dana-farber.org
Pediatric Brain Tumor Foundation: www.pbtfus.org

PEDIATRIC STROKE (See also: Stroke; Epilepsy; Cerebral Palsy) Pediatric Stroke Network: www.pediatricstrokenetwork.com

PITUITARY DISORDERS (See also: Brain Tumor)
Pituitary Network Association: www.pituitary.org

POLIO/POST-POLIO SYNDROME
Post-Polio Health International: www.post-polio.org

POSTPARTUM DEPRESSION
Postpartum Support International: www.postpartum.net

POST-TRAUMATIC STRESS DISORDER
National Center for PTSD: www.ncptsd.org

PRADER-WILLI SYNDROME
Prader-Willi Syndrome Association USA: www.pwsausa.org

PROGRESSIVE SUPRANUCLEAR PALSY
Cure PSP Foundation: www.curepsp.org

PSEUDOTUMOR CEREBRI (See: Rare Disorders)

RARE DISORDERS
National Organization for Rare Disorders (NORD): www.rarediseases.org

REFLEX SYMPATHETIC DYSTROPHY SYNDROME (RSDS)
RSDS Association: www.rsds.org

RESTLESS LEGS SYNDROME
RLS Foundation: www.rls.org

RETT SYNDROME
International Rett Syndrome Association: www.rettsyndrome.org

REYE'S SYNDROME
National Reye's Syndrome Foundation: www.reyessyndrome.org

SCHIZOPHRENIA
National Alliance on Mental Illness (NAMI): www.nami.org

SJOGREN'S SYNDROME
Sjogren's Syndrome Foundation: www.sjogrens.org

SLEEP DISORDERS (See also: Narcolepsy)
American Sleep Apnea Association: www.sleepapnea.org
National Sleep Foundation: www.sleepfoundation.org

SMELL AND TASTE (CHEMOSENSORY) DISORDERS (See Government Resources: National Institute on Deafness and Other Communication Disorders)

SOTOS SYNDROME
Sotos Syndrome Support Association: www.sotossyndrome.org

SPASMODIC DYSPHONIA National Spasmodic Dysphonia Association: www.dysphonia.org

SPASMODIC TORTICOLLIS
National Spasmodic Torticollis Association: www.torticollis.org

SPINA BIFIDA
Spina Bifida Association of America: www.spinabifidaassociation.org

SPINAL CORD INJURY (See also: Disability and Rehabilitation)
National Spinal Cord Injury Association: www.spinalcord.org Paralyzed Veterans of America: www.pva.org
Spinal Cord Injury Network International: www.spinalcordinjury.org

SPINAL MUSCULAR ATROPHY (See also: Muscular Dystrophy) Families of Spinal Muscular Atrophy: www.curesma.com

SPINE-RELATED INJURY/BACK PAIN
North American Spine Society: www.spine.org

STROKE
American Stroke Association: www.strokeassociation.org
National Stroke Association: www.stroke.org

STURGE-WEBER DISEASE
The Sturge-Weber Foundation: www.sturge-weber.com

STUTTERING
National Stuttering Association: www.westutter.org
The Stuttering Foundation: www.stutteringhelp.org

Free Stock Photo

SYRINGOMYELIA
American Syringomyelia & Chiari Alliance Project: www.asap.org

TAY-SACHS DISEASE
National Tay-Sachs and Allied Diseases Association: www.ntsad.org

TINNITUS
American Tinnitus Association: www.ata.org

TOURETTE SYNDROME
Tourette Syndrome Association: www.tsa-usa.org

TRIGEMINAL NEURALGIA
TNA The Facial Pain Association: www.tna-support.org

TUBEROUS SCLEROSIS
Tuberous Sclerosis Alliance: www.tsalliance.org

VESTIBULAR DISORDERS
Vestibular Disorders Association: www.vestibular.org

VON HIPPEL-LINDAU SYNDROME
VHL Family Alliance: www.vhl.org

WILLIAMS SYNDROME
Williams Syndrome Association: www.williams-syndrome.org

WILSON'S SYNDROME
Wilson's Disease Association: www.wilsonsdisease.org

Also available

VISIT www.healingthebrainbook.com

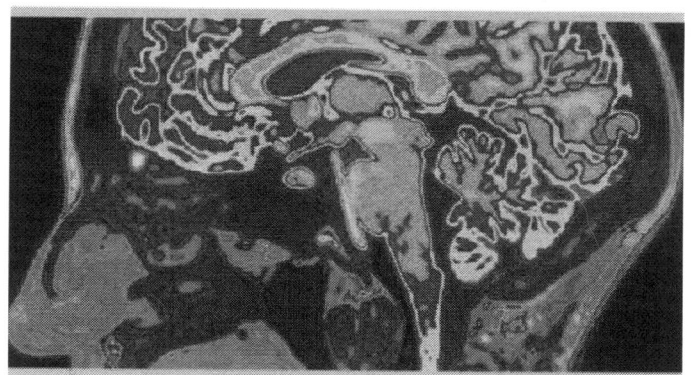

Printed in Great Britain
by Amazon